Springer Theses

Recognizing Outstanding Ph.D. Research

Aims and Scope

The series "Springer Theses" brings together a selection of the very best Ph.D. theses from around the world and across the physical sciences. Nominated and endorsed by two recognized specialists, each published volume has been selected for its scientific excellence and the high impact of its contents for the pertinent field of research. For greater accessibility to non-specialists, the published versions include an extended introduction, as well as a foreword by the student's supervisor explaining the special relevance of the work for the field. As a whole, the series will provide a valuable resource both for newcomers to the research fields described, and for other scientists seeking detailed background information on special questions. Finally, it provides an accredited documentation of the valuable contributions made by today's younger generation of scientists.

Theses are accepted into the series by invited nomination only and must fulfill all of the following criteria

- They must be written in good English.
- The topic should fall within the confines of Chemistry, Physics, Earth Sciences, Engineering and related interdisciplinary fields such as Materials, Nanoscience, Chemical Engineering, Complex Systems and Biophysics.
- The work reported in the thesis must represent a significant scientific advance.
- If the thesis includes previously published material, permission to reproduce this must be gained from the respective copyright holder.
- They must have been examined and passed during the 12 months prior to nomination.
- Each thesis should include a foreword by the supervisor outlining the significance of its content.
- The theses should have a clearly defined structure including an introduction accessible to scientists not expert in that particular field.

More information about this series at http://www.springer.com/series/8790

Saeid Hedayatrasa

Design Optimisation and Validation of Phononic Crystal Plates for Manipulation of Elastodynamic Guided Waves

Doctoral Thesis accepted by
University of South Australia, Adelaide, Australia

Springer

Author
Dr. Saeid Hedayatrasa
School of Engineering
University of South Australia
Adelaide
Australia

Current affiliation:

Department of Materials,
Textiles and Chemical Engineering
Ghent University
Ghent
Belgium

Supervisor
Prof. Kazem Abhary
University of South Australia
Adelaide
Australia

ISSN 2190-5053 ISSN 2190-5061 (electronic)
Springer Theses
ISBN 978-3-319-72958-9 ISBN 978-3-319-72959-6 (eBook)
https://doi.org/10.1007/978-3-319-72959-6

Library of Congress Control Number: 2017962549

Printed on acid-free paper

This Springer imprint is published by Springer Nature
The registered company is Springer International Publishing AG
The registered company address is: Gewerbestrasse 11, 6330 Cham, Switzerland

Supervisor's Foreword

Dynamic loads and associated vibrations and elastic-acoustic waves are inherent in many applications; desirably such as ultrasonic waves used for inspection of human body and wireless communication, or undesirably such as vibrations induced by acoustic-structure-borne sources. Hence, reliable design and optimisation of applied devices and constitutive microstructure is essential for tailored response based on the intended functionality.

Phononic crystals (PhCrs) are acoustic metamaterials (AMMs) with nonhomogeneous periodic layout exhibiting extraordinary characteristics in filtering, resonating and guiding elastic-acoustic waves within particular frequency ranges, called bandgaps. Therefore, maximised bandgap width of PhCrs is desirable in order to expand their phononic controllability over widest frequency range.

The topological optimisation technology is a popular methodology for achieving innovative designs, including PhCrs. Dr. Hedayatrasa's Ph.D. research took the challenge to step into a virgin ground in this field, which is highly commendable given the very competitive environment in topology optimisation for novel PhCrs. It is particularly remarkable that he not only addressed the challenges of new topology optimisation methods for computational design of phononic crystals but also demonstrated the suitability and applicability of the topological designs by experimental validation.

Phononic crystal plates (PhPs) have promising applications in manipulation of guided waves, e.g. design of low loss acoustic devices and built-in acoustic metamaterial. Nonetheless, very limited research works have been reported concerning topology optimisation of PhPs and experimental validation of optimised topologies. Hence, Dr. Hedayatrasa's thesis was particularly dedicated to development of topology optimisation of PhPs for manipulation of guided waves and in this context he made novel contributions in relation to designing and optimising PhCrs in general.

He optimised bi-material PhPs, explored variation of optimised topology versus filling fraction of constituents and introduced and evaluated non-uniform phononic structures with a spatial gradient of filling fraction. Moreover, he optimised porous PhPs and presented the essential role of incorporating the effective stiffness in optimisation of porous PhPs. Among advanced heterogeneous metamaterials, porous type has well-known advantages due to its relative ease of production, light weight and absence of interfacial defects. Obtaining a feasible porous phononic material and with acceptable rigidity is quite challenging. Dr. Hedayatrasa developed robust topology optimisation approaches to overcome this challenge. He produced selected optimised topologies and experimentally demonstrated their promising bandgap structural performance. Also, he optimised porous PhPs for maximised tunability via mechanical deformation and introduced novel PhPs with promising deformation induced switchability/tunablity performance or deformation insensitivity.

Adelaide, Australia
October 2017

Prof. Kazem Abhary

Parts of this thesis have been published in the following documents:

Journals

Saeid Hedayatrasa, Mathias Kersemans, Kazem Abhary, Mohammad Uddin and Wim Van Paepegem, 'Optimisation and experimental validation of stiff porous phononic plates for widest complete bandgap of mixed fundamental guided wave modes', *Mechanical Systems and Signal Processing*, Volume 98, Pages 786–801 (January 2018)

Saeid Hedayatrasa, Mathias Kersemans, Kazem Abhary, Mohammad Uddin, James K. Guest and Wim Van Paepegem, 'Maximizing bandgap width and in-plane stiffness of porous phononic plates for tailoring flexural guided waves: topology optimisation and experimental validation', *Mechanics of Materials*, Volume 105, Pages 188–203 (February 2017)

Saeid Hedayatrasa, Kazem Abhary, Mohammad Uddin and James Guest, 'Optimal design of tunable phononic bandgap plates under equibiaxial stretch', *Smart Materials and Structures*, Volume 25 (5), Pages 05525 (March 2016)

Saeid Hedayatrasa, Kazem Abhary, Mohammad Uddin and Ching-Tai Ng, 'Optimum design of phononic crystal perforated plate structures for widest bandgap of fundamental guided wave modes and maximised in-plane stiffness', *The Mechanics and Physics of Solids*, Volume 89, Pages 31–58 (January 2016)

Saeid Hedayatrasa, Kazem Abhary and Mohammad Uddin, 'Numerical study and topology optimisation of 1D periodic bi-material phononic crystal plates for bandgaps of low order Lamb waves', *Ultrasonics*, Volume 57, Pages 104–124 (March 2015)

Conferences

Saeid Hedayatrasa, Kazem Abhary, Mohammad Uddin, 'On topology optimisation of acoustic metamaterial lattices for locally resonant bandgaps of flexural waves', *The Second Australasian Acoustical Societies' Conference*, Brisbane, Australia, November 2016

Saeid Hedayatrasa, Kazem Abhary, Mohammad Uddin and Ching-Tai Ng, 'Novel Approach in Topology Optimisation of Porous Plate Structures for Phononic Bandgaps of Flexural Waves', *11th world congress of Structural and Multidisciplinary optimisation*, Sydney, Australia, June 2015

Acknowledgements

First and foremost, I would like to express my sincere thanks and gratitude to my principal supervisor Professor Kazem Abhary for his invaluable support throughout my Ph.D. pursuit from scholarship application to the finalisation of my thesis. I appreciate all his contribution of time and ideas to make my Ph.D. research productive and innovative.

My co-supervisor Dr. Mohammad Uddin is also thankfully appreciated for his continuous guidance and insightful advices to enhance the quality of my Ph.D.

The University of South Australia and School of Engineering, in particular, are sincerely appreciated for providing me with University President Ph.D. Scholarship and a comfortable and quiet environment to do my research, and also for awarding me funds to attend international conferences and visit other research groups.

My thanks is extended to the Mechanics of Materials and Structures (MMS) research group at Ghent University, especially Professor Wim Van Paepepegem, Professor Mathias Kersemans and Dr. Nicolas Lammens, who kindly provided me the great opportunity of joining their team and fully supported me in experimental evaluation of my results. I also thank Cléante Dierickx (Odisee) and Kurt Van Houtte (UGent) for manufacturing the phononic plates through water jetting and laser cutting.

Professor James K. Guest, the director of Johns Hopkins Topology Optimisation Group at Johns Hopkins University, is gratefully appreciated for his insightful ideas and invaluable contributions in my Ph.D. research.

The continuous support and advices of Dr. Ching-Tai Ng at University of Adelaide are also deeply appreciated.

I would also like to take this opportunity to thank Professor Tinh Q. Bui at Tokyo Institute of Technology and Dr. Abbas Saboktakin Rizi at University of Quebec for their great help and support in pursuing my research goals.

Last but not least, I would like to thank my wife, my son, my sister and my brother for their love and spiritual support, and express my deep appreciation to my parents for their immense support and kind encouragements throughout my life and education.

Contents

Abbreviations

A	Asymmetric Lamb wave mode
AMM	Acoustic metamaterial
BESO	Bidirectional evolutionary structural optimisation
DIC	Digital image correlation
EFC	Equal frequency contour
FEM	Finite element method
FFT	Fast Fourier transformation
GA	Genetic algorithm
LDV	Laser Doppler vibrometer
LRAB	Locally resonant acoustic bandgap
NSGA	Non-dominated sorting genetic algorithm
PhCr	Phononic crystal
PhP	Phononic crystal plate
PMMA	Plexiglas
PZT	Piezoelectric transducer
RBW	Relative bandgap width
S	Symmetric Lamb wave mode
SH	Shear horizontal wave mode
SHA	Asymmetric shear horizontal guided wave mode
SHS	Symmetric shear horizontal guided wave mode

Symbols

A	Lattice periodicity vector
$\mathbf{A}_b$	Lattice spanning vector of unit-cell's opposing boundaries
a	Unit-cell width
b	Geometrical feature of prescribed topology for bi-material 1D phononic crystal plate
b_s	Stress ratio (ratio of shear to normal stress)
B	Strain–displacement matrix of finite element model
C_{ij}	Elastic property tensor constant
C_g	Group velocity of guided waves
C_{ph}	Phase velocity of guided waves
C_p	Longitudinal wave speed in a thin waveguide
C_s	Shear wave speed
c_b	Perturbation buckling load factor
C	Right Cauchy–Green deformation tensor
$\tilde{\mathbf{C}}$	Volume preservative component of right Cauchy–Green deformation tensor
D	Material property matrix
$\mathbf{D}_\mathbf{e}$	Effective homogenised material property matrix
D	Geometrical feature of prescribed topology for bi-material 1D phononic crystal plate
d	Width of phononic structure
E	Elastic modulus
E_s	Elastic modulus of solid background or stiff inclusion
E_r	Relative elastic modulus of optimised porous topology to pure solid material
E_e	Effective homogenised in-plane elastic modulus of plate unit-cell
E_{FEM}	Effective elastic modulus of phononic unit-cell based on finite element analysis of tensile test

$\overline{E}_{ext}$	Experimentally defined effective elastic modulus of phononic unit-cell through measurements of mechanical extensometers and by assumption of uniform load distribution between unit-cells
$\overline{E}_{DIC}$	Experimentally defined effective elastic modulus of phononic unit-cell through measurements of 3D strereovision and by assumption of uniform load distribution between unit-cells
$\overline{E}_{FEM}$	Effective elastic modulus of phononic unit-cell based on finite element analysis of tensile test and by assumption of uniform load distribution between unit-cells
$\mathbf{e}$	Linear (rotation dependent) component of Green–Lagrange strain tensor
F_1	First objective function
F_2	Second objective function
$\mathbf{F}$	Deformation gradient tensor
$\mathbf{F}_d$	Maximum applied deformation gradient tensor for bandgap tuning
f	Frequency
f_d	Dimensionless frequency
$\mathbf{f}$	Load vector
$\mathbf{f}_{int}$	Vector of internal loads of finite element model
$\mathbf{f}_{cr}$	Vector of critical buckling load
$\mathbf{f}_{ext}$	Vector of external loads of finite element model
$\mathbf{f}_d$	Vector of maximum applied external load for bandgap tuning
G	Shear Modulus
G_e	Effective homogenised in-plane shear modulus of plate unit-cell
G_s	Shear modulus of solid background or stiff inclusion
h	Unit-cell height
$\mathbf{I}$	Identity matrix
i	Imaginary unit number $\sqrt{-1}$
I_1, I_2, I_3	Stretch invariants
J_m	Strain saturation constant defining the stress stiffening nonlinearity of Gent hyperelastic model
k	Circular wave number
k_d	Dimensionless wave number
k_b	Bulk modulus
$\mathbf{k}$	Circular wave vector
$\mathbf{k}_{\mathbf{Re}}$	Real wave vector
$\mathbf{k}_{\mathbf{Im}}$	Imaginary wave vector
$\mathbf{K}$	Stiffness matrix of finite element model
$\mathbf{K}_t$	Tangential stiffness matrix of deformed finite element model
$\mathbf{K}_{\mathbf{p}}$	Linear perturbation stress stiffness matrix
$\mathbf{K}_s$	Stress stiffness matrix of deformed finite element model
$\mathbf{L}$	Lagrangian multiplier matrix
$\mathbf{M}$	Mass matrix of finite element model
$\mathbf{N}$	Matrix of shape functions of finite element model

n_a	Number of additional modal branches included in calculation of total relative and gap width after deformation
n_m	Number of modal branches of interest included in calculation of total relative bandgap width
n_k	Number of wave vectors discretely searched over Brillouin zone border to define modal band structure
P_e	Effective homogenised properties
P_s	Material property of solid background
P_v	Material property assigned to void section
$\mathbf{q}$	Vector of nodal displacements
r_p	Projection radius of mapping design variable domain into topology domain
$\mathbf{R}$	Pure rotation component of deformation gradient tensor
$\mathbf{S}$	Second Piola–Kirchhoff (2PK) stress
t	Time
t_c	Time at the centre of Gaussian envelope
$\mathbf{t}$	Vector of traction forces
u	displacement in x-axis
$\mathbf{u}$	Displacement vector
$\mathbf{u_{x-}}$	Displacement vector of unit-cell boundary node on negative side of x-axis
$\mathbf{u_{y-}}$	Displacement vector of unit-cell boundary node on negative side of y-axis
$\mathbf{u_{x+}}$	Displacement vector of unit-cell boundary node on positive side of x-axis
$\mathbf{u_{y+}}$	Displacement vector of unit-cell boundary node on positive side of y-axis
$\mathbf{u_p}$	Periodic component of displacement vector in lattice structure
$\mathbf{U}$	Pure deformation component of deformation gradient tensor
U	Strain energy
v	displacement in y-axis
v_f	Filling (or volume) fraction
V	Volume
w	displacement in z-axis
W	Strain energy density
$\mathbf{X}$	Current material coordination system after deformation
$\mathbf{x}$	Material coordination system (before deformation)
α_1, α_2, α_3	Vertical components of in-plane normal, in-plane shear and anti-plane shear elastic wave numbers respectively
$\boldsymbol{\alpha}$	Nonlinear (rotation altering) component of Green–Lagrange strain tensor
β	Logarithmic frequency sweep coefficient
γ	Shear strain
γ_G	Gabor wavelet coefficient

γ	Green–Lagrange strain tensor
Γ	Brillouin zone point standing for in-plane wave vector $\{\pi/a \quad 0\}$
$\boldsymbol{\varepsilon}$	True strain tensor
$\overline{\boldsymbol{\varepsilon}}$	Average macromechanical strain of unit-cell by boundary displacements
ε	Normal Strain
ζ	Wavelength
η	Buckling eigenvalue
λ	First Lame elastic constant
λ_1, λ_2, λ_3	Eigenvalue of right Cauchy–Green deformation tensor
μ	Second Lame elastic constant
M	Brillouin zone point standing for in-plane wave vector $\{\pi/a \quad 0\}$
v	Poisson's ratio
v_s	Poisson's ratio of solid background or stiff inclusion
v_e	Effective homogenised in-plane Poisson's ratio of plate unit-cell
$\overline{v}_{ext}$	Experimentally defined effective Poisson's ratio of phononic unit-cell through measurements of mechanical extensometers and by assumption of uniform load distribution between unit-cells
$\overline{v}_{DIC}$	Experimentally defined effective Poisson's ratio of phononic unit-cell through measurements of 3D strereovision and by assumption of uniform load distribution between unit-cells
$\overline{v}_{FEM}$	Effective Poisson's ratio of phononic unit-cell based on finite element analysis of tensile test and by assumption of uniform load distribution between unit-cells
ξ	Macromechanical stretch
ξ_d	Maximum applied tensile stretch for bandgap tuning
ρ	Mass density
ρ_s	Mass density of solid matrix material or stiff inclusion
σ	Normal Stress
$\boldsymbol{\sigma}$	True (Cauchy) stress tensor
$\overline{\boldsymbol{\sigma}}$	Average macromechanical stress on unit-cell's boundary
τ	Shear stress
$\boldsymbol{\Phi}$	Binary vector of design variables defining the bi-material or solid-void) topology
X	Brillouin zone point standing for in-plane wave vector $\{\pi/a \quad 0\}$
ψ	Gabor mother wavelet function
ω	Angular frequency
ω_d	Dimensionless angular frequency
ω_R	Maximum bandgap frequency of undeformed unit-cell within modal branches of interest
Ω_F	Relative bandgap width based on modal frequencies
Ω_E	Relative bandgap width based on eigenvalues

Chapter 1
Background and Research Scope

1.1 Introduction

Different materials exhibit distinct properties due to their particular atomic structure. This fact besides advent of modern and precise manufacturing technologies, e.g. additive manufacturing (3D Printing) and laser writing, have inspired design of novel heterogeneous materials-structures with complex shape and microstructure. The engineered microstructure fashioned from conventional materials can produce desired effective properties in a larger scale. Such composite materials whose effective properties are derived from and are controlled actively or passively, by their underlying design are called Functional Materials. If the designed microstructural constitution of functional material leads to extraordinary properties not found in the nature, e.g. negative refraction index or negative Poisson's ratio, then it is called metamaterial.

A functional structure may be produced by non-uniform and non-periodic distribution of constitutive material(s) in the design domain if desired. However, the microstructure of a representative volume element (unit-cell) is generally designed to produce desired functionality at macroscale through a perfectly periodic lattice structure. Furthermore, a multifunctional structure may be designed by spatial gradient of unit-cell's microstructure to make multiscale functionality.

Dynamic loads and associated vibroacoustic waves and vibrations are inherent to many applications. The load could be (i) a disturbance like impact loads of crashes or environmental noise, (ii) an operational load like vibrations induced on aircrafts wings by engine and turbulent flow or (iii) could have a desirable nature like ultrasound waves used for inspection of human body or structures. This necessitates appropriate design and optimisation of applied structures and constitutive materials for tailored vibroacoustic behaviour based on the intended functionality, from small ultrasonic transducers to large automotive parts. This thesis is concerned with topology optimisation of phononic acoustic metamaterial lattices for maximised vibroacoustic controllability as discussed in the following section.

S. Hedayatrasa, *Design Optimisation and Validation of Phononic Crystal Plates for Manipulation of Elastodynamic Guided Waves*, Springer Theses,
https://doi.org/10.1007/978-3-319-72959-6_1

1.2 Acoustic Metamaterials (AMMs)

Acoustic metamaterials (AMMs) are functional materials with extraordinary manipulation capabilities on mechanical vibrations. AMMs control acoustic and elastodynamic wave propagation through three principal mechanisms: (i) phononic scattering of wave at the interface of heterogeneities, (ii) local resonance of microstructural features, and (iii) spatial gradient of effective macromechanical properties leading to variation of refraction index. Acoustic bandgaps are a prominent type of AMMs with promising functionalities in attenuation, resonation, guiding and steering of vibroacoustic waves. Unlike compliant damping materials which dissipate the wave energy, phononic acoustic bandgaps reflect a resonated wave energy, and locally resonant acoustic bandgaps trap the wave energy through local resonances, while maintaining the load bearing capacity.

1.2.1 Phononic Acoustic Bandgaps

Phononic acoustic bandgaps, so called phononic crystals (PhCr), are AMMs with promising manipulation capabilities on propagation of vibroacoustic waves (Deymier 2011) resembling photonic crystals affecting electromagnetic waves. PhCrs are indeed heterogeneous materials produced by periodic modulation of acoustic impedance in a lattice structure through either integration of two or more contrasting materials, or making void inclusions in a single material. The phononic lattice periodicity can be 1D, 2D or 3D to make unidirectional, in-plane omnidirectional or 3D omnidirectional bandgaps respectively.

Figure 1.1 shows the Eusebio Sempere's sculpture which is an example of 2D phononic crystal on display at the Juan March Foundation in Madrid. It is made by a square 10 × 10cm lattice of hollow steel cylinders each 2.9 cm in diameter. In 1995 researchers at the Materials Science Institute of Madrid showed that the sculpture strongly attenuates acoustic waves at certain frequencies. The main feature of PhCrs making them known as acoustic bandgap materials is the existence of such frequency ranges over which propagation of vibroacoustic waves is prohibited. This phenomenon is caused by in-phase reflection and superposition of waves at the interface of periodic heterogeneities i.e. Bragg scatterings. Additionally the internal reflections of scattering inclusions may also contribute to the bandgap called Mie scatterings (Olsson Iii and El-Kady 2009). Therefore the efficiency of PhCr is wavelength dependent and the frequency range of bandgap is limited by its lattice periodicity constant or in other words its feature size.

For example the modal band structure and phononic bandgap of a perforated aluminium plate of thickness 25 mm with 50 × 50 mm square lattice of circular holes is calculated and shown in Fig. 1.2 for different wave vectors in the lattice plane. The higher shaded area highlights the complete bandgap over which there is absolutely no modal response for all wave propagation directions. However, partial

Fig. 1.1 Eusebio Sempere's sculpture Órgano, on display at the Juan March Foundation in Madrid, as an example of phononic crystals (PhCrs)

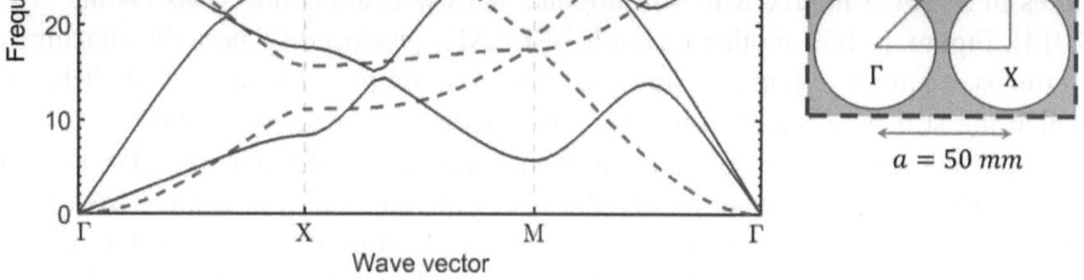

Fig. 1.2 Modal band structure and phononic bandgap of an aluminium plate of thickness 25 mm with 50×50 mm square lattice of circular perforations each 45 mm in diameter

bandgaps may exist with bandgap efficiency in particular wave vectors only, like the lower frequency range shaded in the band structure depicted in Fig. 1.2 The solid and dash lines separate symmetric and asymmetric plate modes to be described later on in Sect. 1.3.

Due to the strong wavelength scale heterogeneity of PhCrs, they possess other surprising wave manipulation properties. Introducing any kind of defect in phononic lattice e.g. by altering the features of a few adjacent cells, leads to advent of local resonance modes within phononic bandgap frequency. This capability is used to trap and guide waves inside defects at desired frequencies within bandgap (Olsson Iii and El-Kady 2009; Zhu and Semperlotti 2013). Moreover, studying the equal frequency contours of PhCrs show flat and concave wave fronts around bandgap frequency applicable for self-collimation and steering of waves (Lin et al. 2009; Wang et al. 2013). Consequently, it is worthwhile to adjust the phononic properties so that the width and location of bandgap is optimised for application of interest. The efficiency of PhCr is principally defined by the features of its irreducible representative element (Unit-cell) i.e. its geometry, topology and composition.

1.2.2 Locally Resonant Acoustic Bandgaps

Acoustic bandgaps can also be produced through specific periodic features that attenuate vibrations through destructive internal oscillations at their resonation frequencies. In fact, the resonant inclusions/attachments act as internal oscillators and cause very large or negative effective dynamic mass within bandgap frequencies. Such AMMs are called locally resonant acoustic bandgaps (LRAB). LRABs can be produced by periodic insertion of resonating inclusions in a background material. The inclusions may be produced by insertion of a stiff-dense material in a soft-compliant intermediate material relative to the background material. LRABs can also be produced by periodic attachment of resonating stubs to a background material. Figure 1.3a shows a LRAB produced by insertion of cylindrical copper cores in a rubber matrix with 4 orthogonal narrow beam connections (Wang et al. 2014). Figure 1.3b is another example of LRABs produced by periodic attachment of tungsten-silicon cylindrical stubs on an aluminium plate (Assouar et al. 2012), in which the stiff tungsten stacked over soft silicon base is the oscillator.

LRABs have the same wave manipulation properties of PhCrs on vibroacoustic waves. The main advantage of LRABs over PhCrs is their operational frequency. PhCrs attenuate waves by successive in-phase reflection of wave at the interface of periodic scatters. Consequently the phononic bandgap operates at frequencies with wavelengths lower or in the order of lattice periodicity.

It is well proven that longer wavelength of lower frequencies does not interact with phononic scatters. However, the operational frequency of LRABs depends on the resonation frequency of locally resonant features. Therefore for specified lattice periodicity of the same background material a LRAB can manipulate much lower frequencies as compared to a PhCr.

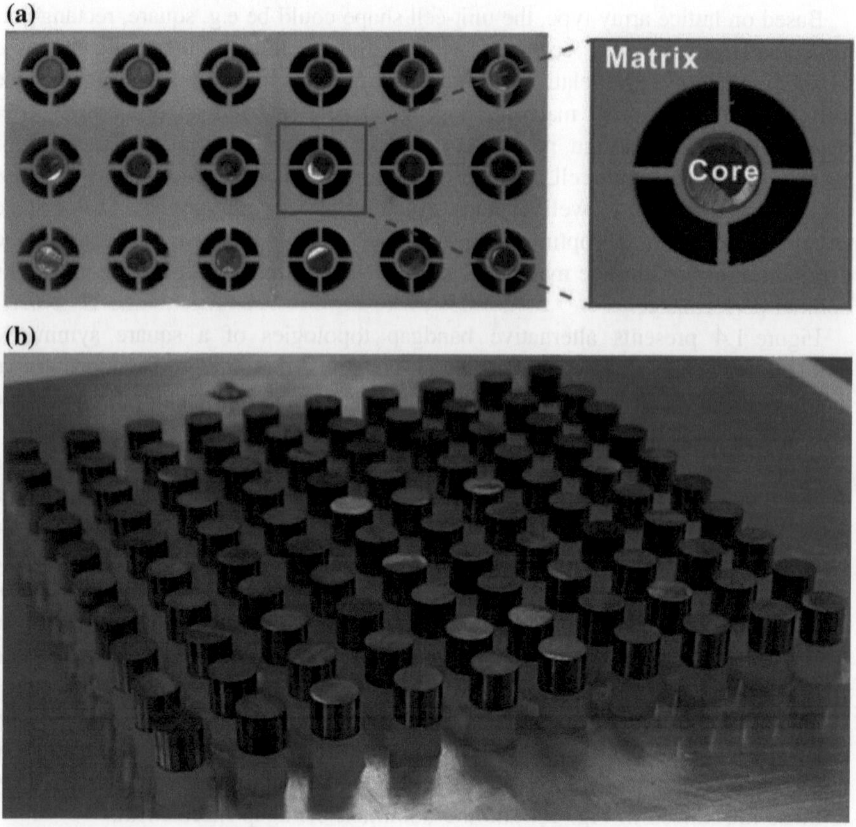

Fig. 1.3 Examples of LRABs **a** copper cylindrical cores inserted in silicon rubber matrix (Wang et al. 2014), **b** thin aluminium plate with tungsten/silicon rubber cylindrical stubs (Assouar et al. 2012)

1.3 Design and Optimisation of (Phononic) Acoustic Bandgaps

With regards to the promising characteristics of acoustic bandgaps, their optimal design for maximised bandgap efficiency is desirable. Due to periodic construction of acoustic bandgaps, their efficiency is principally defined by the features of irreducible representative element (unit-cell) including its geometry, topology and material composition. The main characteristics of bandgap to be addressed are bandgap width, frequency level, modal location and effective direction or isotropy. So the basic features of unit-cell can be tailored to achieve particular bandgap properties of interest. Generally widest bandgap at lowest frequency range or in other words maximum relative bandgap width (RBW) is desired. Therefore widest bandgap frequency range is achieved through smallest unit-cell size compared with the wavelength.

Based on lattice array type, the unit-cell shape could be e.g. square, rectangular, hexagonal or a ring section, which highly affects the efficiency of bandgap. Moreover the relative mechanical properties and so acoustic impedance mismatch of constitutive materials have great role in bandgap properties. After specifying primary design parameters (the shape and material constitution) of acoustic bandgap's unit-cell, its microstructural topology predominantly defines bandgap performance as well as macrostructural effective properties. At this stage an appropriate topology optimisation technique enables the designer to define the distribution of constitutive material(s) in the design domain (unit-cell) to ensure its optimal performance.

Figure 1.4 presents alternative bandgap topologies of a square symmetric unit-cell with two distinctively different constitutive martials. The constitutive material displayed in dark colour is stiff and dense (e.g. copper) and the other one displayed in light colour is relatively light and compliant (e.g. rubber).

Bragg scattering of wave at the interface of a stiff scatter in a compliant matrix as depicted in Fig. 1.4a can open a phononic bandgap. Localised Mie resonances of a compliant scatter inside a stiff matrix as shown in Fig. 1.4b can also open a bandgap at relatively low resonance frequency of scatters (Hsu and Wu 2007; Li et al. 2004; Wang et al. 2004). The LRAB can be further enhanced by addition of an oscillatory stiff-dense core inside the compliant scatter as depicted in Fig. 1.4c which reduces the local resonance frequency and widens the bandgap width (Hedayatrasa et al. 2016).

Implemented topology optimisation strategy and definition of its bandgap objective function and also assumed relative stiffness and density of constitutive materials have essential role in the resemblance of an optimised bandgap topology to those introduced in Fig. 1.4 and the dominant bandgap mechanism.

If the unit-cell's constitution is single material (i.e. porous) and the compliant domain is replaced with void, then among alternative topologies of Fig. 1.4 only the one shown in Fig. 1.4b is feasible. Hence, Mie local resonance of compliant core is invalid and the porous unit-cell may have bandgap properties due to interfacial

(a) Bragg scattering (Compliant matrix) (b) Mie local resonance (Stiff matrix) (c) Oscillatory core and enhanced local resonance

Fig. 1.4 Schematic presentation of alternative bi-material topologies for a square symmetric unit-cell, and dominant bandgap mechanisms

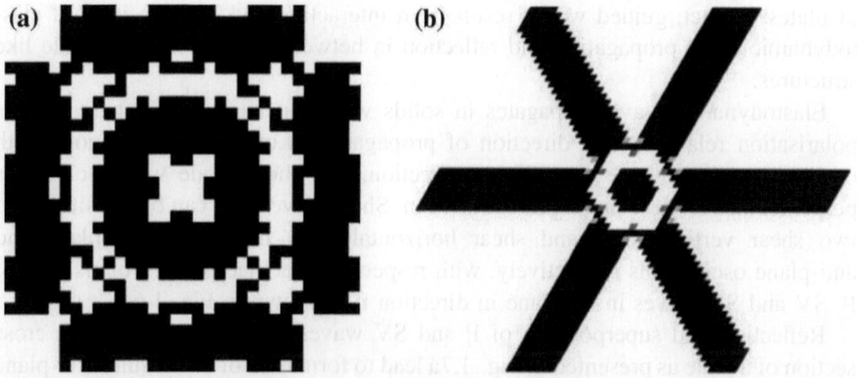

Fig. 1.5 Optimised phononic crystal plate unit-cells with locally resonant features, **a** in this study, and **b** by Halkjær et al. (2006)

reflections and resonances of its solid matrix (e.g. Fig. 1.2). Nonetheless, when performing topology optimisation of a porous unit-cell for maximised bandgap efficiency, topologies with infeasible disconnected features (and virtual locally resonant bandgap performance) like Fig. 1.4c may still appear depending on the robustness of optimisation strategy in treating the void areas.

Figure 1.5 presents two examples of optimised porous PhCr unit-cells having dominant locally resonant features. The square symmetric topology shown in Fig. 1.5a was achieved in this study for maximised RBW of low order guided waves which contains a central resonating inclusion with weak hinge connection to its surrounding structure. The hexagonal topology given in Fig. 1.5b is another example obtained by Halkjær et al. (2006) for maximised RBW of guided waves. This topology also has a resonating solid core which is artificially connected to the surrounding frame through assumed compliant material considered for porosities.

In general, the overall bandgap characteristic of topologically optimised unit-cells (which normally have irregular construction) cannot simply be explained by Bragg scattering alone which by definition occurs in phononic bandgaps. The unit-cell's topology may have several segments which all contribute to shifting and broadening of bandgap through phononic scattering and local resonance. However, the topologies studied in this thesis have dominant phononic bandgaps and termed as PhCr, as common in the literature.

1.4 Bandgaps of Guided Waves in Plate Structures

AMMs can be designed to manipulate bulk waves in an infinite structure, surface wave in a semi-infinite structures or guided waves. Guided waves are structural waves confined by traction free surfaces of thin walled structures like parallel faces of plates. In fact, guided waves result from interaction and superposition of elastodynamic wave propagation and reflection in between thin passages of plate like structures.

Elastodynamic wave propagates in solids with two basic modes based on its polarisation related to the direction of propagation: Longitudinal (P) mode with oscillations along wave propagation direction, and shear mode with oscillations perpendicular to wave propagation direction. Shear wave itself can be classified into two shear vertical (SV) and shear horizontal (SH) modes with in-plane and anti-plane oscillations respectively, with respect to specified plane. Polarisation of P, SV and SH waves in xy-plane in direction **n** is shown in Fig. 1.6.

Reflection and superposition of P and SV waves propagating along the cross section of a plate as presented in Fig. 1.7a lead to formation of two resultant in-plane modes called Lamb waves: symmetric Lamb mode (S) and asymmetric Lamb mode (A) with respect to the mid-plane of plate's thickness, schematically shown in Fig. 1.7b, c respectively. Likewise, superposition of SH waves produces resultant symmetric (SHS) and asymmetric (SHA) shear horizontal guided wave modes.

Guided waves are generally dispersive and their propagation speed is a function of frequency at wavelengths larger than the plate's thickness. Figure 1.8 shows calculated dispersion curves of aluminium plate and frequency dependency of wave speed for various wave modes and orders.

Thin wall elements are fundamental building components of many advanced structures and controlling their vibroacoustic behaviour is of great importance. Phononic scattering can be implemented in design of vibration-less structure at desired frequency ranges. For example phononic structures can be used for stiff attachment of a vibrating device to a substrate while destructing transmission of vibroacoustic waves, or supporting a delicate device like gyroscope and immunising it to the surrounding structural vibrations.

Guided waves travel long distances in plate structures with low loss which makes them ideal for production of low loss AMM resonators, filters and waveguides (Lin et al. 2014; Mohammadi 2010). Low loss transmission and dispersive characteristic of guided waves is also advantageous for structural health monitoring (SHM) and non-destructive evaluation (NDE) purposes (Su and Ye 2009; Veidt et al. 2008).

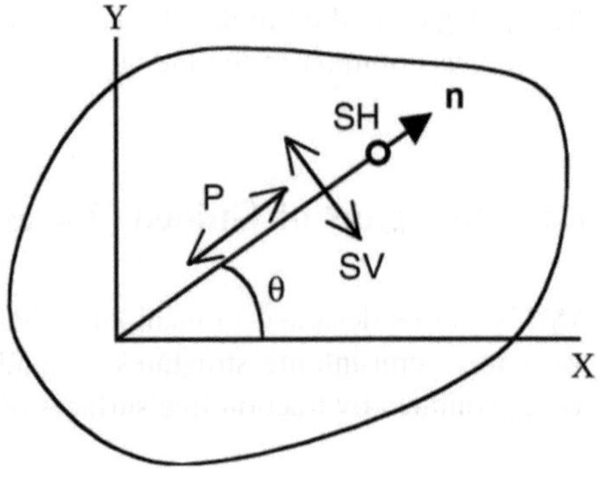

Fig. 1.6 Polarisation of P, SV and SH waves propagation in xy-plane in direction **n** (Kundu 2004)

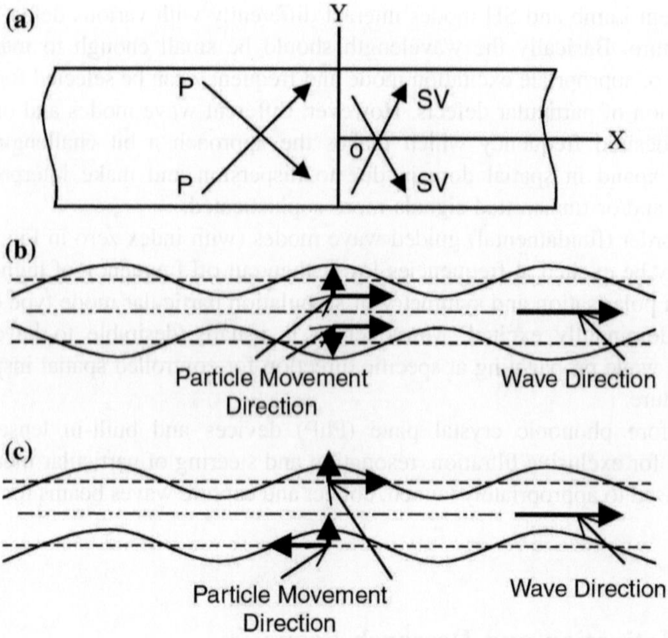

Fig. 1.7 **a** Reflection and superposition of P and SV waves propagating within cross section of a plate and formation of **b** symmetric and **c** asymmetric Lamb modes (Kundu 2004)

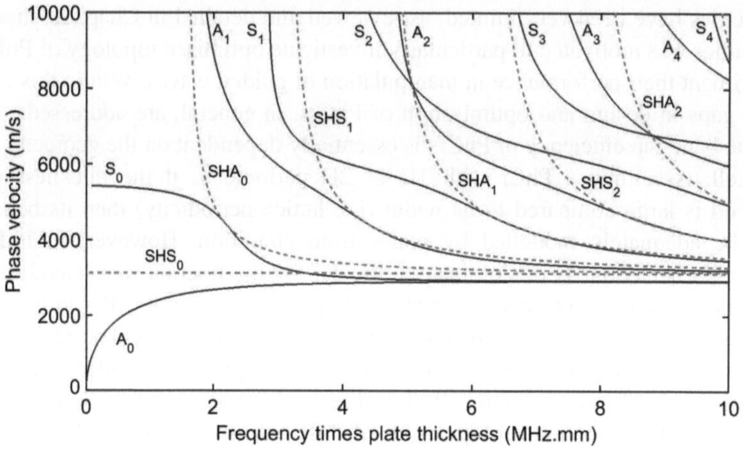

Fig. 1.8 Dispersion curves of guided waves in aluminium plate and frequency dependency of wave speed for different wave modes and orders

Different Lamb and SH modes interact differently with various defect modes of the structure. Basically the wavelength should be small enough to interact with defects. So, appropriate excitation mode and frequency can be selected for adequate qualification of particular defects. However, different wave modes and orders may exist at desired frequency which makes the approach a bit challenging. Wave packets expand in spatial domain due to dispersion and make interpretation of reflected and/or transmitted signals more sophisticated.

First order (fundamental) guided wave modes (with index zero in Fig. 1.8) may preferably be excited at frequencies lower than cut off frequency of higher modes. Based on polarisation and symmetry of stimulation particular mode type of interest can be dominantly excited. Moreover, it is usually desirable to have a beam detective wave propagating at specific direction for controlled spatial inspection of the structure.

Therefore phononic crystal plate (PhP) devices and built-in lenses can be designed for exclusive filtration, resonation and steering of particular modes which may be used to appropriately launch, collect and capture waves beams for SHM and NDE.

1.5 Motivation and Research Scope

In spite of promising application and efficiency of PhPs, the research and studies in relation to their design optimisation and experimental validation of optimised topologies have been very limited, as reviewed and detailed in Chap. 2. Therefore, the author was motivated to particularly investigate optimised topology of PhPs and experiment their performance in manipulation of guided waves. Within this context other gaps in design and optimisation of PhCrs, in general, are addressed.

The bandgap efficiency of PhCrs is essentially dependent on the geometry of the unit-cell. Assuming a PhCr with 1D or 2D periodicity, if the thickness of the unit-cell is large compared to its width (i.e. lattice periodicity) then its behaviour may be adequately modelled by plane-strain condition. However, in PhPs the unit-cell's behaviour is predominantly affected by the traction free lateral faces of its finite thickness. For specified plate thickness, various unit-cell widths may be designed to achieve different bandgap efficiencies. Different thicknesses also may be considered for specified unit-cell width to adjust the bandgap properties. Obviously, the thicker the unit-cell, the stiffer and the heavier it is. So it is of great value to study the optimised topologies of different aspect ratios and their relative bandgap efficiencies.

PhPs for bandgaps of asymmetric guided wave modes have been solely explored in earlier studies. However, it is also worthwhile to investigate the topology of PhPs for maximised RBW of symmetric modes as well as complete bandgap of mixed guided wave modes. Complete bandgap of guided waves can be used as vibration-less supporting structure of delicate dynamic device like gyroscope for immunity to noise sources. Guided waves can be also manipulated through exceptional flat and concave equal frequency contours produced near the bandgap frequency of corresponding mode, for self-collimation and wave steering purposes.

If the bandgap is exclusive to either symmetric or asymmetric modes, then the modes can be handled individually. Allowed guided wave modes can be almost purely launched as the banned modes are filtered within the corresponding exclusive bandgaps. Bandgap mode on the other hand can be trapped or guided within appropriate defects in PhP lattice. So, detective wave beams of symmetric or asymmetric modes can be individually excited, steered and captured through PhP elements with promising application in structural health monitoring.

The acoustic mismatch of constitutive materials in a PhCr is vital for opening bandgaps. High contrast of constitutive materials (i.e. elastic modulus and density) is desired for magnifying interfacial reflections, and so the weight and overall stiffness of PhCr is highly dependent on the filling fraction of constitutive materials. Hence, it is worthy to study the variation of optimised topology and its bandgap efficiency with respect to filling fraction. Moreover, the effective bandgap efficiency of gradient PhCrs produced through variation of e.g. filling fraction of unit-cells have not been addressed. Such gradient PhCr can be optimised in both unit-cell and structure scale for multiscale functionality.

Manufacturability of PhCrs is also an essential issue which should be taken into account. Porous PhCrs produced through perforation of a uniform solid background are relatively easy to produce in which the porous interfaces act as strong reflectors of wave energy. In optimisation of porous PhCrs it is vital to appropriately handle the discontinuity of design domain and also ensure about the feasibility and stiffness of optimised topology. However, treatment of discontinuities and the contribution of stiffness in optimisation of porous PhCrs have not been addressed in the literature.

Designing PhCrs with tunability/switchability performance through an external stimulation is a new topic which is in its infancy. An approach in tuning the bandgap is to deform the phononic lattice under an external load which introduces tunability through deformation-induced gradient of both topology and stiffness. Deformation-induced degradation of bandgap in PhCrs with prescribed topology has been well investigated. However, there has been no study reported in relation to optimality of PhCrs for tailored bandgap tunability performance through deformation.

Consequently, this thesis basically explain the topology optimisation of PhPs, particularly porous type, to address aforementioned gaps in literature. The scope of this research is detailed in Chap. 2.

1.6 Concluding Remarks

In this chapter, the characteristics of AMMs in manipulation of elastodynamic and acoustic waves was explained. Phononic and locally resonant acoustic bandgaps were discussed and the role of topology optimisation for enhancing the bandgap efficiency of PhCrs was discussed. Application of PhPs in manipulation of guided waves in thin-walled structures for design of low loss vibroacoustic devices and structural health monitoring was explained. The lacks of literature in topology optimisation of PhCrs and specifically PhPs was briefly mentioned and the need for developing topology optimisation in this area was argued. Finally, the research scope targeting topology optimisation of PhPs was introduced.

References

Assouar, M. B., Senesi, M., Oudich, M., Ruzzene, M., & Hou, Z. (2012). Broadband plate-type acoustic metamaterial for low-frequency sound attenuation. *Applied Physics Letters, 101*(17), 173505.

Deymier, P. (2011). *Acoustic metamaterials and phononic crystals*. Berlin: Springer.

Halkjær, S., Sigmund, O., & Jensen, J. S. (2006). Maximizing band gaps in plate structures. *Structural and Multidisciplinary Optimization, 32*(4), 263–275.

Hedayatrasa, S., Abhary, K., & Uddin, M. (2016). On topology optimization of acoustic metamaterial lattices for locally resonant bandgaps of flexural waves. *The 2nd Australasian Acoustical Societies' Conference.*

Hsu, J.-C., & Wu, T.-T. (2007, May 14). Lamb waves in binary locally resonant phononic plates with two-dimensional lattices, *Applied Physics Letters, 90*(20), pp. 201904–201903.

Kundu, T. (2004). *Ultrasonic nondestructive evaluation—Engineering and biological material characterization*. Boca Raton: CRC Press.

Li, J., & Chan, C. (2004). Double-negative acoustic metamaterial. *Physical Review E, 70*(5), 055602.

Lin, C.-M., Hsu, J.-C., Senesky, D. G., & Pisano, A. P. (2014) Anchor loss reduction in ALN Lamb wave resonators using phononic crystal strip tethers. In *Frequency Control Symposium (FCS), 2014 IEEE International*, pp. 1–5.

Lin, S.-C. S., Huang, T. J., Sun, J.-H., & Wu, T.-T. (2009). Gradient-index phononic crystals. *Physical Review B, 79*(9), 094302.

Mohammadi, S. (2010) Phononic band gap micro/nano-mechanical structures for communications and sensing applications. Georgia Institute of Technology.

Olsson Iii, R. H., & El-Kady, I. F. (2009). Microfabricated phononic crystal devices and applications. *Measurement Science & Technology, 20*(1), 012002.

Su, Z., & Ye, L. (2009). *Identification of damage using lamb waves: From fundamentals to applications*. Berlin: Springer.

Veidt, M., Ng, C.-T., Hames, S., & Wattinger, T. (2008). Imaging laminar damage in plates using Lamb wave beamforming. *Advanced Materials Research, 47,* 666–669.

Wang, P., Casadei, F., Shan, S., Weaver, J. C., & Bertoldi, K. (2014). Harnessing buckling to design tunable locally resonant acoustic metamaterials. *Physical Review Letters, 113*(1), 014301.

Wang, P., Shim, J., & Bertoldi, K. (2013). Effects of geometric and material nonlinearities on tunable band gaps and low-frequency directionality of phononic crystals. *Physical Review B, 88*(1), 014304.

Wang, G., Wen, X., Wen, J., Shao, L., & Liu, Y. (2004). Two-dimensional locally resonant phononic crystals with binary structures. *Physical Review Letters, 93*(15), 154302.

Zhu, H., & Semperlotti, F. (2013). Metamaterial based embedded acoustic filters for structural applications. *AIP Advances, 3*(9), 092121.

Chapter 2
Literature Review and Research Objectives

2.1 Introduction

The characteristics of AMMs and PhCrs in particular and their design optimisation were discussed in Chap. 1. The lacks of literature in topology optimisation of PhCrs was briefly mentioned and the need for developing topology optimisation in this area was argued. The application of PhPs for manipulation of guided waves was explained and the research scope targeting topology optimisation of PhPs was introduced.

The aim of this chapter is to present and discuss the aforementioned gaps in detail. The existing literature concerning optimisation of AMMs, particularly PhCrs and PhPs, are categorised and explained. The research problem and objectives are then introduced based on the shortages of literature. Finally the thesis structure and the contents of thesis chapters are detailed.

2.2 Optimisation of AMMs and Objectives

The extraordinary behaviour of metamaterials and structures is achieved through engineered heterogeneous layout of one, two or multiple materials in specified design domain. Hence, topology optimisation has been competently employed to exploit the best material lay out to tailor the effective behaviour of metamaterials and maximise their desired performance. Examples are metamaterials with widest photonic bandgap (Kao et al. 2005; Meng et al. 2015), negative/zero/extreme effective thermal expansion (Wang et al. 2016a; Wang et al. 2016b) and negative Poisson's ratio (Wang et al. 2015; Wang et al. 2014).

Optimum design of AMMs has been also widely studied in literature in order to achieve desired functionality in micro (unit-cell) and/or macro (lattice-structure) scales. Various optimisation objectives have been considered for controlling the

S. Hedayatrasa, *Design Optimisation and Validation of Phononic Crystal Plates for Manipulation of Elastodynamic Guided Waves*, Springer Theses,
https://doi.org/10.1007/978-3-319-72959-6_2

vibrational and wave propagation behaviour under transient vibroacoustic loading. In this section the objectives concerning optimisation of AMMs are fundamentally categorised and corresponding research examples are explained.

2.2.1 Optimisation of Finite Non-periodic AMM Structure

A finite domain is considered for the AMM structure and optimum composition of constitutive material(s) inside the domain is obtained for specified objectives. The finite structure may be independent or may be an element of larger structure.

Rupp et al. (2007) implemented gradient based algorithm for topology optimisation of finite 3D AMM aluminium-silicon structure to design filters and wave guide devices for bulk waves and more specifically surface waves. The optimised bulk wave filter and surface wave guide are shown in Fig. 2.1a and b respectively. In Fig. 2.1, the volumes enclosed by grey surfaces indicate the presence of silicon, and white surfaces indicate aluminium.

Larsen et al. (2009) implemented gradient based topology optimisation for designing bi-material Mindlin's plate structures with supressed vibration or wave guiding characteristic at desired excitation frequencies. They considered square 50×50 cm and 3 mm thick steel-polycarbonate plates with maximum allowed fraction of steel of 25%. Figure 2.2a and b show achieved optimised topologies for minimised vibration at $\omega = 2790\,\mathrm{rad/s}$ and maximised horizontal energy transport from left edge to specified area in the middle-right respectively, with simply supported boundaries.

Du and Olhoff (2010) minimised the sound pressure emitted by a plate structure into a reference plane in the acoustic medium as depicted in Fig. 2.3a, where the plate is subjected to a uniform acoustic pressure at specified frequency. Bi-material square plate with width to thickness of 0.02 was assumed and the optimised

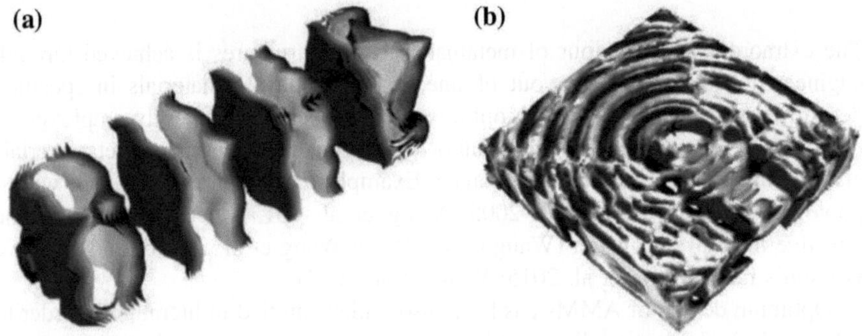

Fig. 2.1 Optimised finite 3D AMM structure for **a** filtration of bulk waves at specified frequency from one end to another, and **b** focusing normal distributed surface load on top-center to the specified length of top edge at prescribed frequency (Rupp et al. 2007)

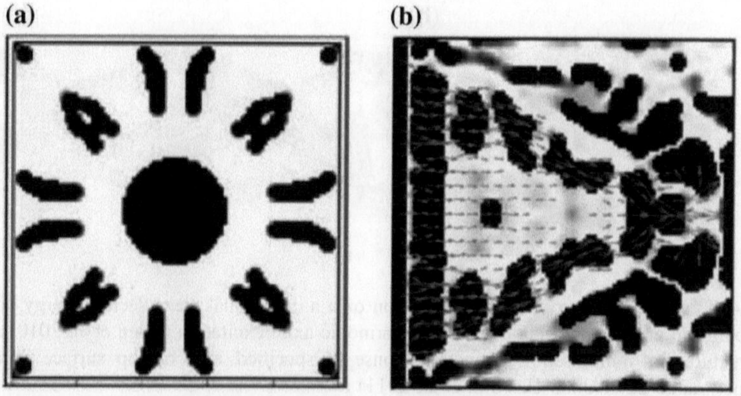

Fig. 2.2 Optimisation of finite bi-material square plate structure (Larsen et al. 2009) for **a** minimised vibration at $\omega = 2790\,\mathrm{rad/s}$, and **b** maximised horizontal energy transport from left edge to specified area in the middle-right; black area stands for steel

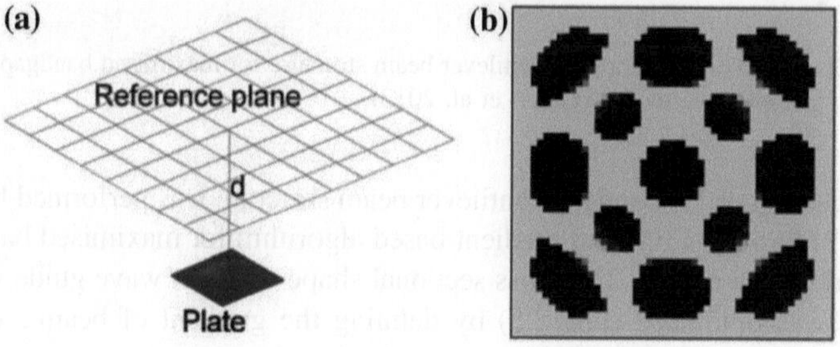

Fig. 2.3 **a** vibrating plate structure and reference plane in the acoustic medium, **b** optimised bi-material plate for minimised sound radiation into the reference plane at $\omega = 10,000\,\mathrm{rad/s}$ (Du and Olhoff 2010)

topology was achieved through gradient based approach. Figure 2.2b shows the optimised topology for minimised sound radiation at $\omega = 10,000\,\mathrm{rad/s}$ having clamped boundaries.

Level-set structural topology optimisation was employed by (Chen et al. 2010) for topology optimisation of single and multimaterial vibration based piezoelectric energy harvesters. As an example, the topology of a single material cylindrical energy harvester was optimised (Fig. 2.4a) for maximised energy conversion subjected to harmonic axial load at specified frequency.

Shu et al. (2011) also implemented level-set based topology optimisation to minimise the frequency response of finite structures at specified points or surfaces subjected to predefined excitation frequency (Fig. 2.4b). Modelling and sensitivity analysis is performed by extended finite element method (X-FEM) and void areas are removed from analysis.

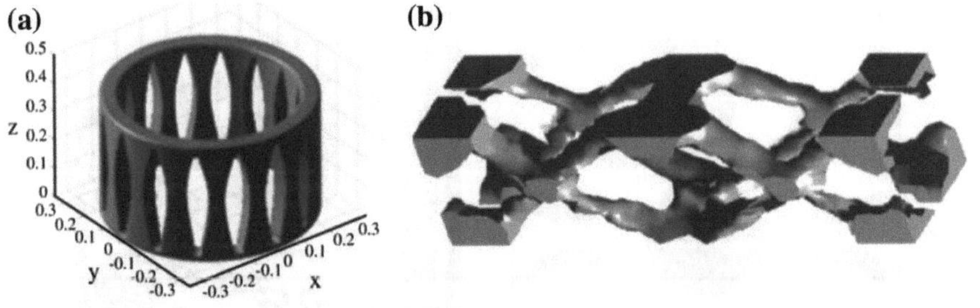

Fig. 2.4 Level-set based topology optimisation of **a** a cylindrical piezoelectric energy harvester for maximized energy conversion at 250 Hz harmonic axial excitation (Chen et al. 2010) and **b** a 3D structure for minimised frequency response at specified area of top surface within the frequency range 2000–3000 Hz (Shu et al. 2011)

Fig. 2.5 Optimised cross section of cantilever beam structure for maximised bandgap of selected asymmetric guided wave modes (Olhoff et al. 2012)

Optimisation of 1D bandgap cantilever beam structure was performed by (Olhoff et al. 2012) through FEM and gradient based algorithm for maximised bandgaps of asymmetric wave modes. The cross sectional shape of beam wave guide with finite thickness was optimised (Fig. 2.5) by defining the gradient of beam's cross sectional area (with the same geometry) over its length.

2.2.2 *Optimisation of Finite AMM Lattice Structure*

Finite AMM lattice structure can also be designed where the vibroacoustic behaviour of desired finite domain is controlled by its perfect or gradient periodic microstructure.

El-Sabbagh et al. (2008) proposed gradient based topology optimisation of Mindlin's plates with periodic microstructure for maximising fundamental frequency. The periodic unit-cell was characterised by a uniform grid of different elements with various thicknesses as design variable. The plate structure was formed by a periodic arrangement of unit-cell to be optimised (Fig. 2.6), and then the whole structure was analysed micromechanically. Principally direct micromechanical analysis of a finite structure provides the capability of applying realistic boundary conditions and analysing the low to high frequency transient response of structure in an accurate manner. But the imposed computational cost degrades its efficiency for the case of large problems.

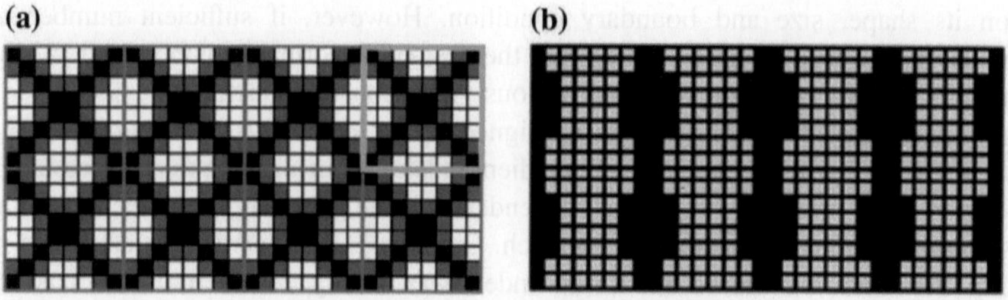

Fig. 2.6 **a** finite rectangular 300 × 150 mm plate with 4 × 2 square unit-cells, and **b** optimised topology for maximised fundamental frequency at overall filling fraction 50% and varying thickness from 0.5 to 5 mm (El-Sabbagh et al. 2008)

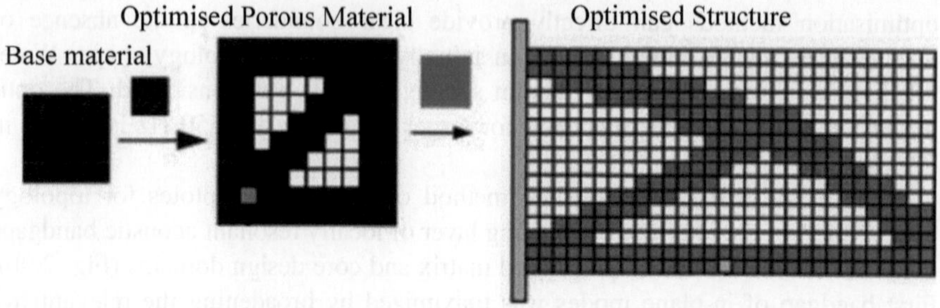

Fig. 2.7 Multiscale design of an ultralight structure for maximum fundamental frequency with porous microstructure using homogenisation (Niu et al. 2009)

Niu et al. (2009) developed a methodology to find ultralight lattice structure with maximum fundamental frequency by utilising topology optimisation at both micro (unit-cell) and macro (structure) scales. Sequential linear programming with explicit sensitivity analysis is implemented as optimisation strategy. Micro and macro densities are introduced as independent continuous variables and concurrently optimised using homogenisation theory without iteration between two scales. The micro density varies from base material to zero (void), and macro density varies from homogenised density to zero (Fig. 2.7). No hierarchy is considered for material scale to achieve a homogeneous structure.

2.2.3 *Optimisation of AMM Unit-Cell Under Perfect Periodicity*

AMM lattice with perfect periodicity can also be designed by merely considering the unit-cell scale. Appropriate boundary conditions are required to prescribe perfect periodicity leading to desired macromechanical effective properties when performing as a cell of whole structure. The behaviour of lattice structure depends on its shape, size and boundary condition. However, if sufficient number of unit-cells are present in the lattice then the desired performance is obtained. AMM unit-cells have been optimised for various vibroacoustic objectives.

An optimised AMM lens was designed by Li et al. (2012) using GA. The objective was to design an acoustic gradient index lens in air, with maximised index of refraction, minimised frequency dependence of the material properties, as well as minimised acoustic impedance mismatch. Optimised 2D acoustic lens as shown in Fig. 2.8 with maximised refraction index was finally obtained at the cost of increased frequency dependence.

Lu et al. (2013) proposed a level set based optimisation method for design of AMMs with minimised dynamic bulk modulus at chosen target frequencies. This leads to negative bulk modulus and consequently causes wave propagation suppression at specified frequency. Level set-based topology (and practically shape) optimisation methods can directly provide clear boundaries in the absence of intermediate pseudo domains from an initially prescribed topology shape. Water filled rubber AMM with 20 × 20 mm square unit-cell was considered. The optimised topology of rubber inclusion for target frequency of 1550 Hz is shown in Fig. 2.9a.

Yang et al. (2016) implemented method of moving asymptotes for topology optimisation of the intermediate coating layer of locally resonant acoustic bandgaps (LRABs) confined between prescribed matrix and core design domains (Fig. 2.9b). First bandgap of in-plane modes was maximized by broadening the relevant frequency range with negative dynamic mass density characteristic.

Design and optimisation of PhCr unit-cells have been widely studied in literature for desired phononic wave manipulation. Park et al. (2015) achieved optimised square bi-material 2D PhCrs for self-collimation of SH bulk waves along a target direction through a gradient based method. Optimised topology for self-collimation

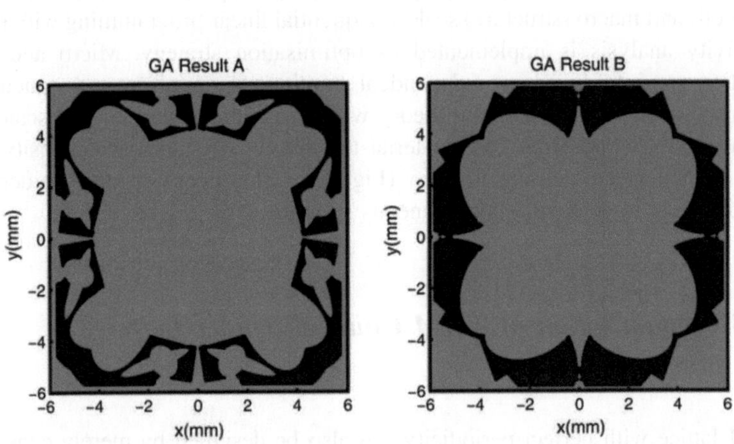

Fig. 2.8 GA optimised AMM lens designs (Li et al. 2012), black area stands for solid background and blue area for void

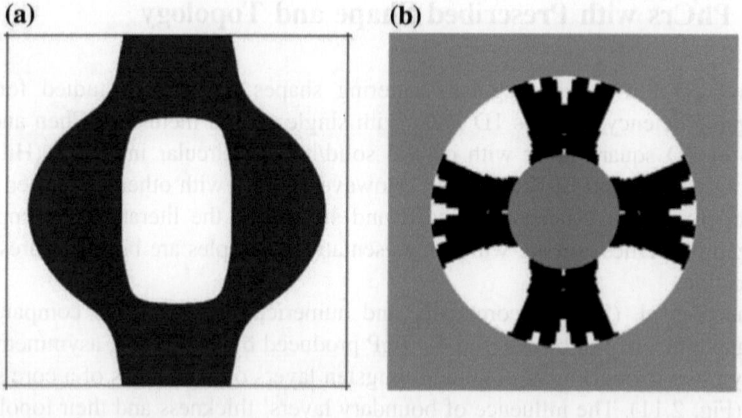

Fig. 2.9 Topology optimisation of locally resonant bandgaps **a** water filled rubber inclusion for minimised negative bulk modulus at target frequency of 1550 Hz (Lu et al. 2013) and **b** optimised topology of intermediate coating layer of locally resonant acoustic bandgap for minimized dynamic mass density at specified frequency above lowest bandgap (Yang et al. 2016)

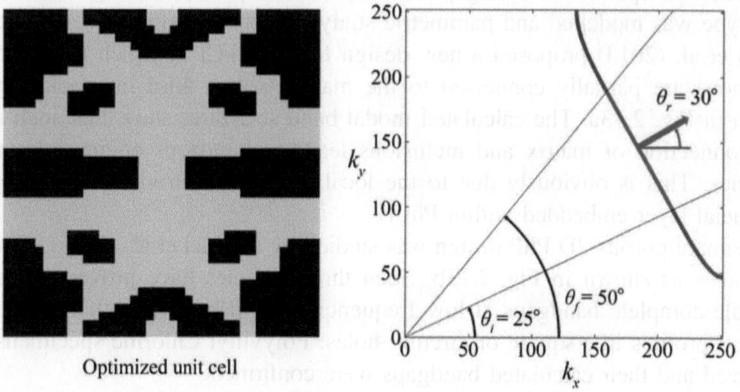

Fig. 2.10 Optimised bi-material square 2D PhCr for maximised self-collimation of bulk SH waves along specified angle $\theta = 30^{\circ}$ (Park et al. 2015); Blue area stands for silicon inclusion and yellow area for epoxy background

along desired angle $\theta = 30^{\circ}$ is shown in Fig. 2.10 in which a central flat wave front is evident perpendicular to this target angle.

However, the main characteristic of PhCrs is the existence of acoustic bandgaps over which propagation of vibroacoustic wave is banned. Many of earlier studies are dedicated to improved design and optimisation of PhCrs for maximised bandgap efficiency which will be addressed in detail in the following sections.

2.3 PhCrs with Prescribed Shape and Topology

1D and 2D PhCrs with regular scattering shapes have been studied for their bandgap efficiency, such as 1D PhCr with single central inclusion (Chen and Han 2010) or 2D square PhCr with central solid/hollow circular inclusion (Hu et al. 2004; Olsson Iii and El-Kady 2009). However, PhCrs with other prescribed shape and topologies have been introduced and studied in the literature for improved bandgap properties, among which representative examples are briefly addressed in this section.

Huang et al. (2014) theoretically and numerically studied and compared the bandgap efficiency of bi-material 1D PhP produced by symmetric, asymmetric and anti-symmetric attachment of silicon-tungsten layers on both sides of a core silicon plate (Fig. 2.11). The influence of boundary layers' thickness and their topological distribution on bandgaps was studied and significant distinction was observed between their efficiency.

Yilmaz et al. (2007) introduced a novel hexagonal 2D PhCr (Fig. 2.12) with embedded amplification mechanisms by which the dynamic inertia of lattice is controlled to open wide bandgaps at very low frequencies. Simple mass-spring prototype was modelled and parametric study on bandgap efficiency was studied.

Yu et al. (2013) proposed a new design for 2D PhCr in which the cylindrical inclusions are partially connected to the matrix with a third interfacial layer as shown in Fig. 2.13a. The calculated modal band structures show that such imperfect connection of matrix and inclusions leads to bandgaps at much lower frequencies. This is obviously due to the local resonance introduced by the partial interfacial layer embedded within PhCr.

A square porous 2D PhP design was studied by Miniaci et al. (2015) with cross link holes as shown in Fig. 2.13b. Such through holes have proved to nucleate multiple complete bandgaps at low frequencies compared with other regular perforation profiles like square or circular holes. Polyvinyl Chloride specimens were produced and their calculated bandgaps were confirmed.

Although efficient PhCrs can be achieved by parametric study of predefined shape of topology inclusions, optimum performance can be approached through an

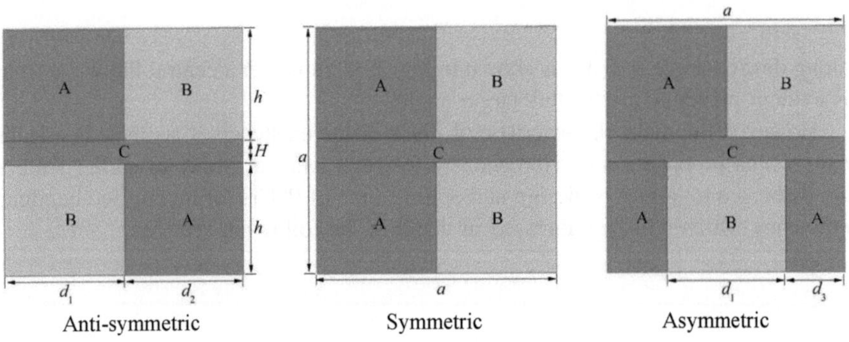

Fig. 2.11 1D PhP with different configurations of boundary layers (Huang et al. 2014)

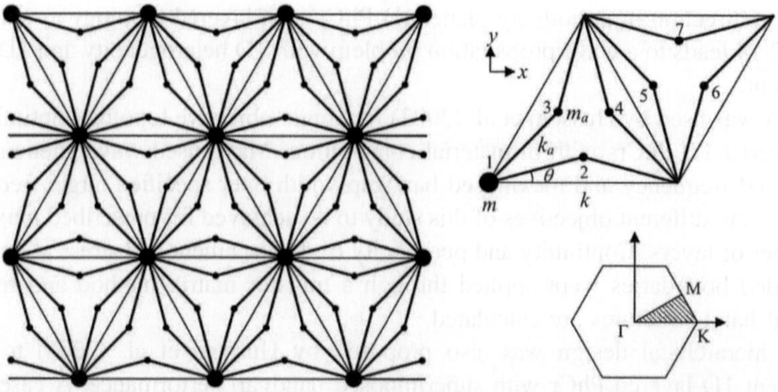

Fig. 2.12 Prototype hexagonal 2D PhCr with internal amplification mechanism (Yilmaz et al. 2007)

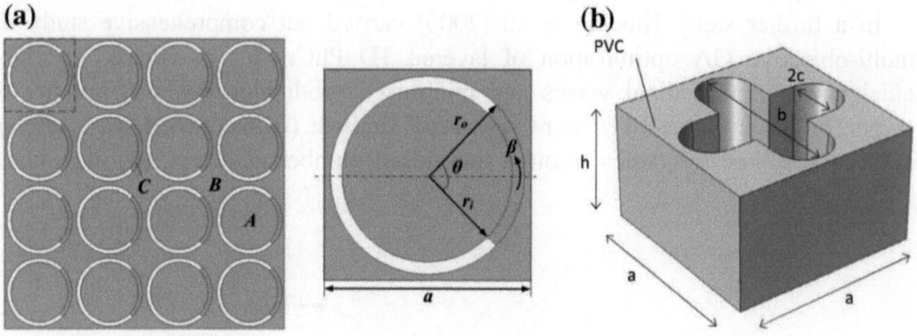

Fig. 2.13 2D PhP designs with **a** imperfect bounding of cylindrical inclusion A in matrix B with partial interfacial layer C leading to locally resonant bandgap at low frequencies (Yu et al. 2013), **b** square porous topology with cross link holes having relatively better bandgap efficiency compared with other regular (e.g. square or circular) perforation profiles (Miniaci et al. 2015)

appropriate topology optimisation procedure. The unit-cell shape and topology constraints are prescribed and the topology is optimised for desired bandgap performance. Following sections are dedicated to discussion of preceding works on topology optimisation of PhCrs.

2.4 Topology Optimisation of Very Thick 1D PhCrs

Many of PhCr topology optimisation studies are concerned with manipulation of bulk waves through a sufficiently thick periodic medium perpendicular to the corresponding lattice periodicity vector. Layered topology with 1D gradient or 2D gradient topology layout may be assumed for unidirectional wave manipulation along a direction in periodicity plane. 1D PhCr with layered topology as shown in Fig. 2.14 leads to a wave propagation problem with 1D heterogeneity and 1D strain gradient.

GA was used by Hussein et al. (2002) for single objective topology optimisation of layered 1D PhCrs with bi-material constitution. Maximised wave attenuation at specified frequency and maximised bandgap width over specified target frequency range were different objectives of this study to be achieved for prescribed maximum number of layers. Continuity and periodicity of displacement and stress at perfectly bounded boundaries were applied through a transfer matrix method and relevant modal band structures are calculated.

A hierarchical design was also proposed by Hussein et al. (2005) to make gradient 1D layered PhCr with superimposed bandgap performance. A case study was presented in which the pass bands of an optimised topology were covered by bandgaps of another subsequent optimised topology. Both topologies are optimised for maximised bandgap over target frequency range and the second one is specifically designed to cover bandgaps of first one (Fig. 2.15).

In a further study Hussein et al. (2006) carried out comprehensive study on multi-objective GA optimisation of layered 1D PhCrs for maximised bandgap efficiency of longitudinal waves and minimised production cost i.e. number of layers. Following objectives were separately studied: (i) maximised wave attenuation at specified frequency through minimised number of layers, (ii) maximised

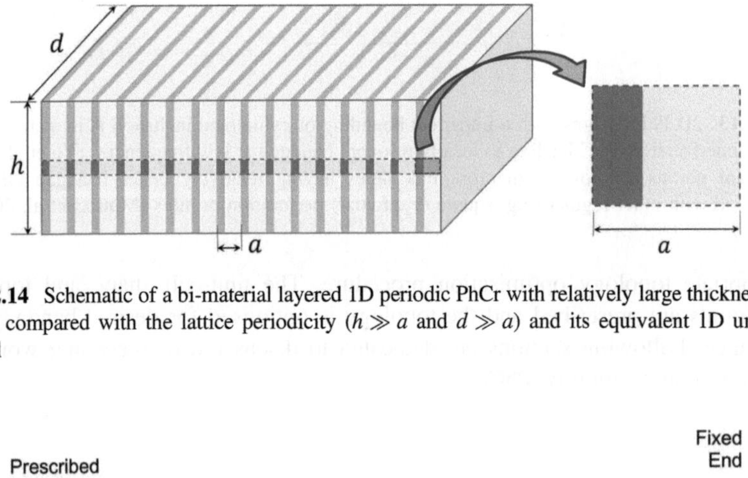

Fig. 2.14 Schematic of a bi-material layered 1D periodic PhCr with relatively large thickness and width compared with the lattice periodicity ($h \gg a$ and $d \gg a$) and its equivalent 1D unit-cell model

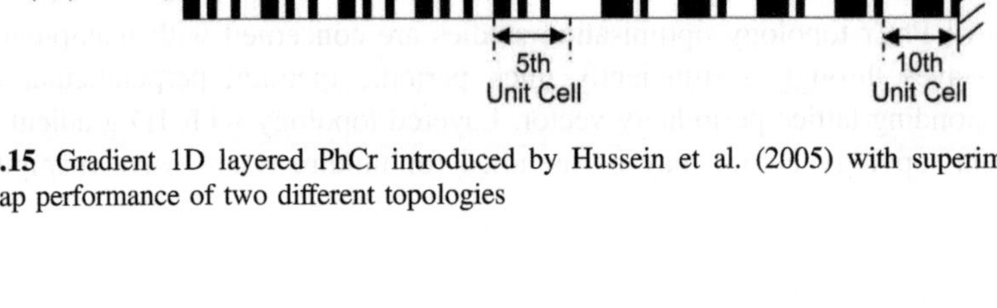

Fig. 2.15 Gradient 1D layered PhCr introduced by Hussein et al. (2005) with superimposed bandgap performance of two different topologies

bandgap width over specified target frequency range through minimised number of layers, and (iii) multiple stop and pass bands at multiple target frequencies.

2.5 Topology Optimisation of Thick 2D PhCrs (Plane-Strain Condition)

2D PhCrs can also be designed for manipulation of waves in all directions in the lattice periodicity plane. If the thickness of PhCr unit-cell is much bigger than its other dimensions then the problem can be reduced to a 2D plane-strain model (Fig. 2.16). The in-plane and anti-plane oscillations are decoupled and may be studied separately or concurrently.

Sigmund and Jensen (2003) conducted comprehensive study on systematic design of PhCrs and finite phononic structures. Bi-material 2D PhCr with fictitious contrasting material properties and square symmetry was assumed and analysed by FEM. Gradient based optimisation was employed to maximise RBW of in-plane and anti-plane modes separately between selected modal branches.

In a further study Halkjær et al. (2005) implemented gradient based optimisation and FEM for maximised RBW of aluminium-Plexiglas (PMMA) 2D PhCr with relaxed symmetry. Square PhCrs with just one axis of symmetry were optimised for unidirectional bandgap of in-plane modes along axis of symmetry. Rhombic unit-cells were also optimised for omnidirectional bandgap of in-plane modes in the phononic lattice plane.

Gazonas et al. (2006) implemented GA optimisation algorithm and FEM for maximised RBW of Ti-SiC and water filled porous ceramic 2D PhCrs. Square and

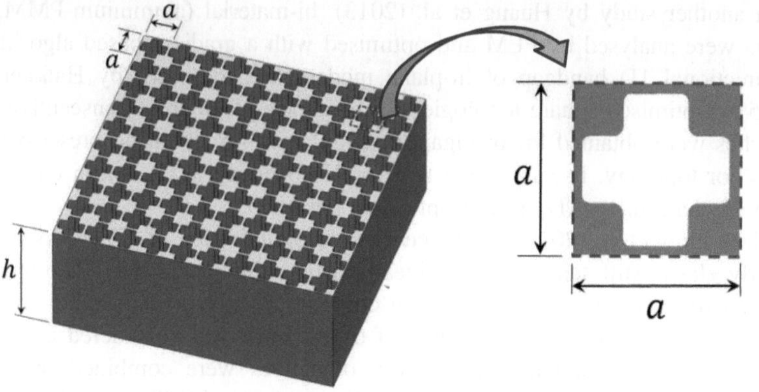

Fig. 2.16 Schematic of an arbitrary 2D periodic porous PhCr with relatively large thickness compared with the lattice periodicity ($h \gg a$) and its equivalent 2D plane-strain unit-cell model

hexagonal unit-cells were considered with no constraint on the symmetry. The topology was discretised by uniform triangular grids in contrary to commonly used square grids.

GA optimisation of 2D bi-material PhCrs was also carried out by Hussein et al. (2007) using FEM. Fictitious stiff and compliant solid materials were assumed and square symmetric PhCrs with maximised total RBW within first 5 modal branches of anti-plane modes were obtained. Parametric study was performed to observe the variation of optimised topologies with respect to various combinations of density and elastic modulus ratios.

Bilal and Hussein (2011) investigated optimum topology of 2D square symmetric PhCrs for maximised RBW of in-plane and anti-plane modes, separately and concurrently, between selected consecutive couple of modal branches. For complete bandgap of concurrently in-plane and anti-plane modes, special lead-follow algorithm was introduced to efficiently treat both objectives in a single objective optimisation algorithm. Single material porous silicon PhCrs were studied and a very compliant material was considered as the void phase of topology. As it is unlikely to have bandgaps opened at the onset of optimisation process, auxiliary terms were added to the bandgap objective. These terms represent a measure for the area between desired consecutive modal branches in the modal band structure to be maximised if no bandgap exists. Scaling factors were fitted in order to prescribe the priority of original bandgap objective to auxiliary terms.

Vatanabea and Silva (2011) employed FEM and optimised bi-material piezo-composite 2D PhCrs through a gradient based approach. Widest bandgap between two adjacent in-plane modal branches was the objective of optimisation. A functionally graded interface was assumed as the intermediate layer of inclusions and applied by a projection approach with different penalisation factors. However post-processing of results confirmed the evident higher bandgap efficiency of topologies with discrete (non-gradient) interface.

In another study by Huang et al. (2013), bi-material (aluminium-PMMA) 2D PhCrs were analysed by FEM and optimised with a gradient based algorithm for unidirectional 1D bandgap of in-plane modes (like the work by Halkjær et al. (2005)). Optimised square topologies for bandgaps of different consecutive modal branches were obtained for bandgap along 90° and 45°, with no prescribed symmetry for topology. It was shown that for 1D bandgap a 2D optimised design is more efficient than a 1D layered optimised design.

Manktelow et al. (2013) employed GA for optimisation of 2D PhCrs made by linearly elastic stiff scatters in nonlinear elastic compliant matrix. The amplitude induced strain sensitivity of dispersion curve i.e. bandgap (or in other words the dependency of bandgap to amplitude of oscillations) was considered as a second objective besides bandgap width. Both objectives were combined in a single objective with appropriate weight factors. It was shown that PhCr can be optimised for maximised bandgap efficiency as well as maximised sensitivity to nonlinear wave propagation. Post processing of optimised topology was performed for feasibility and manufacturability.

A new convex optimisation was introduced by Men et al. (2013) based on linear program and topology optimisation of bi-material (lead-epoxy) 2D PhCrs was carried out. Maximised bandgap width of a few consecutive modal branches of in-plane and anti-plane modes were obtained and the computational efficiency of approach was confirmed.

Dong et al. (2014a) used GA in combination with FEM for optimisation of bi-material square symmetric 2D PhCrs (lead or gold in epoxy matrix) for complete and exclusive bandgaps of in-plane and anti-plane modes. Constrained optimisation with prescribed filling fraction of constituents was also performed to study its effect on optimised topology.

In another work by Dong et al. (2014b) topology of square bi-material (lead-Epoxy) 2D PhCr was optimised with no prescribed symmetry. Multi-elitist GA was employed in conjunction with adaptive fuzzy granulation for faster convergence and reduced computational cost. The results showed enhanced bandgap efficiency of topologies with relaxed symmetry constraints compared with square symmetric ones, for complete and exclusive bandgaps of in-plane and anti-plane modes.

Dong et al. (2014) also performed multi-objective optimisation for maximised bandgaps of silicon porous 2D PhCrs with minimised mass. Non-dominated sorting GA II (NSGA-II) was implemented and geometrical constraint was applied for avoiding mesh dependency of porous topology. Complete bandgap of both in-plane and anti-plane modes as well as their exclusive bandgaps were achieved and compared with earlier work by Bilal and Hussein (2011). For complete bandgaps and exclusive bandgap of in-plane modes the later work by Dong et al. (2014) converged to optimised topologies with thinner connections of scattering segments (i.e. lower structural stiffness) and apparently higher bandgap efficiency. This is obviously due to the different topology filtration approaches and connectivity constraints imposed by these works.

Liu et al. (2014) implemented GA with fast plane wave expansion method for optimisation and evaluation of bi-material (steel-epoxy) 2D PhCrs with square symmetry. A two stage optimisation was considered in which the initially coarse topology is optimised to an acceptable convergence limit and the optimisation is finalised for a refined topology resolution.

Bi-directional evolutionary structural optimisation (BESO) method was also implemented by Fan et al. (2016) for optimisation of 2D bi-material PhCrs. Complete bandgaps of mixed polarisations and exclusive bandgaps of in-plane and anti-plane modes were maximised separately. The developed approach was shown to be computationally efficient due to its fast convergence.

Most recently Yang et al. (2016) implemented level set method for topology optimisation of the intermediate coating layer of locally resonant acoustic bandgaps (LRABs) confined between prescribed matrix and core design domains. First bandgap of in-plane modes was maximized by broadening the relevant frequency range with negative dynamic mass density characteristic.

2.6 Topology Optimisation of PhCr Plates (PhPs)

Many of topology optimisation studies discussed in relation to PhCr lattices have been concerned with bandgaps of in-plane and/or anti-plane bulk waves while only few works have been devoted to bandgaps of guided waves in PhPs. The modal band structure of PhPs is dominantly affected by their finite thickness which is comparable to, and generally smaller than, their lattice periodicity. Figure 2.17a shows a 1D layered PhP with bi-material constitution and its equivalent 2D plane-strain unit-cell model. Figure 2.17b also shows a porous 2D PhP with square symmetry and its 3D unit-cell model.

As discussed earlier in Chap. 1, guided waves are structural waves confined by traction free surfaces of thin walled structures, so called plate waves when guided by parallel faces of plates. In-plane symmetric (S) and anti-plane asymmetric (A) Lamb waves as well as symmetric shear horizontal (SHS) and asymmetric shear horizontal (SHA) in-plane waves are the well-known guided wave modes generated in a plate structure. Special characteristics of guided waves confined by finite thickness of such structures, make them ideal for non-destructive evaluation

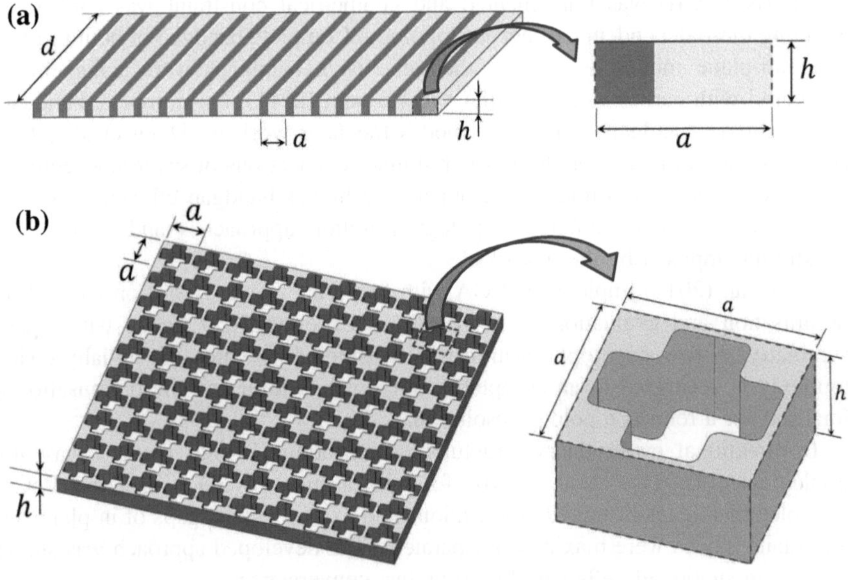

Fig. 2.17 Schematic of **a** a bi-material layered 1D periodic PhP with relatively small thickness and large width compared with the lattice periodicity ($h \leq a$ and $d \gg a$) and its equivalent 2D plane–strain unit-cell model, **b** an arbitrary porous 2D PhP with relatively small thickness compared with the lattice periodicity ($h \leq a$) and relevant unit-cell model

purposes (Su and Ye 2009; Veidt et al. 2008) as well as production of low loss resonators, filters and waveguides (Lin et al. 2014; Mohammadi 2010). The guided wave dispersion in PhP is governed by its transversal anisotropy i.e. plate's thickness in addition to its planar anisotropy measured by lattice periodicity and unit-cell constitution.

The bandgaps of guided waves in 2D PhPs of prescribed topologies have been extensively studied (Charles et al. 2006; Khelif et al. 2006; Pennec et al. 2010; Wu et al. 2011). PhPs produced by either periodic insertion, attachment or perforation of cylindrical inclusions on a background plate have been generally considered in these studies. Phononic lattice with different patterns e.g. square, rectangular or polygonal were considered. The phononic unit-cell itself could contain a specific pattern of circular inclusions like the work by Li, Jianbao, Wang and Zhang (2009) who studied guided wave gaps in PhPs with Archimedean-like tiling.

The first study on topology optimisation of PhPs was carried out by Halkjær et al. (2006). The optimum topology of porous Polycarbonate PhP with rhombic unit-cell and width to thickness ratio of 11 was explored for maximised RBW of first couple of flexural (asymmetric) guided waves. The Mindlin's plate theory was implemented for definition of band structure of bending waves and gradient based optimisation was performed through moving asymptotes method. Since the best topology for maximised RBW (Fig. 2.18a) was infeasible and entirely disconnected, new objective was introduced to conversely search widest bandgap width at higher frequency ranges. However the improved optimised topology as shown in Fig. 2.18b was still infeasible with isolated solid features. This is because of the nature of bandgap objective which tends to make isolated scattering features possibly with local resonance for the sake of low frequency wide bandgaps. The assumed compliant material substituting the porosities of topology artificially enables existence of such optimised topologies. To solve this issue the discontinuities of improved optimised topology (Fig. 2.18b) were locally modified for acceptable stiffness and manufacturability as presented in Fig. 2.18c. This modification led to significant degradation of its RBW value from 0.6 to 0.16. Anyhow, if the infeasible locally resonant features of optimised topologies have been included in calculation of reported RBW of optimised topologies their higher bandgap efficiencies are not actually achievable.

In a further relevant work by Bilal and Hussein (2012) the optimum topology of thin porous silicon PhPs for maximised RBW of basic flexural waves was studied and supreme feasible topology was obtained as compared with that of Halkjær et al. (2006). Mindlin's plate theory was implemented and topology of square symmetric unit-cell was optimised through GA. The discreetness of GA provides the capability to check the feasibility of topology and reject or penalise those with discontinuous features. Thus the topology achieved by this later work as shown in Fig. 2.19 did not need any post-processing for manufacturability.

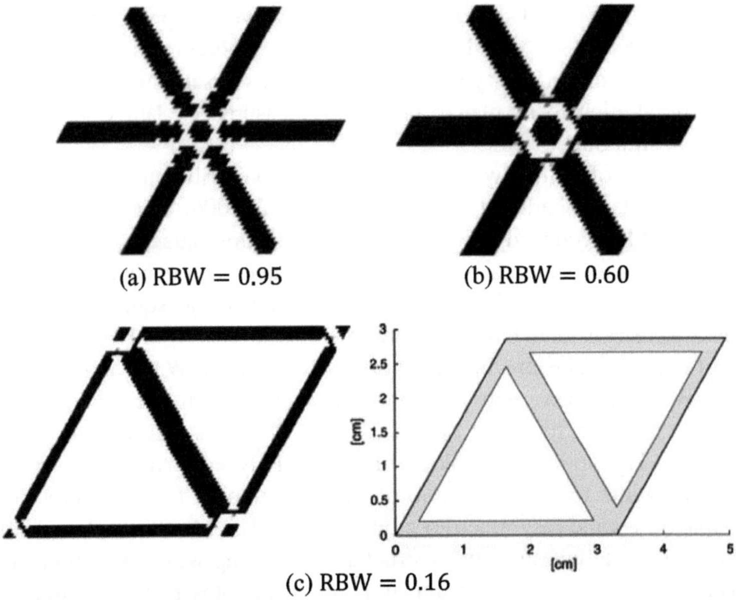

Fig. 2.18 Topology optimisation of porous PhP with rhombic unit-cell shape for maximised bandgap of first couple of asymmetric modal branches (Halkjær et al. 2006) and relevant RBW values, **a** optimised topology for maximised RBW (ratio of bandgap width over mid-gap frequency), **b** optimised topology for maximised product of bandgap width and mid-gap frequency, **c** post processing for feasibility and adequate stiffness

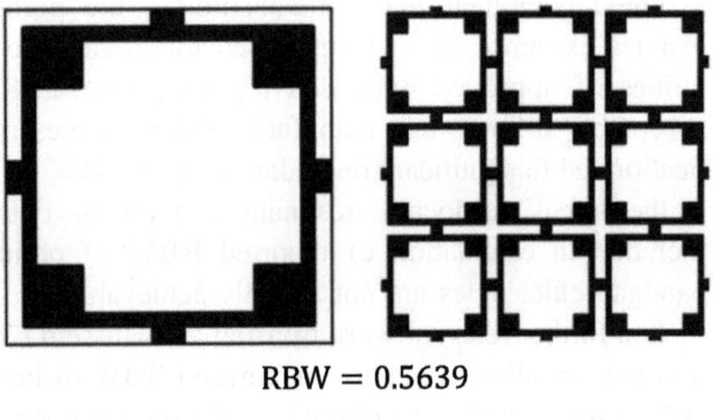

Fig. 2.19 Optimised topology of square symmetric porous PhP for maximised RBW of first couple of asymmetric modal branches by Bilal and Hussein (2012), and its RBW value for the same material properties and aspect ratio considered by Halkjær, Sigmund and Jensen (2006)

2.7 Tunable PhCrs and Topology Optimisation

Due to wide multiscale application of PhCrs it is of great importance to introduce controllable phononic bandgaps to be tailored, degraded or enhanced during their function or with switchable bandgap properties (Matar et al. 2013). For instance an adaptive phononic wave guide can be designed so that its functional frequency or direction can be changed during operation. This can be achieved through use of active features, such as tailored changes in topology and/or material properties. For example active piezoelectric material can be used locally to control the effective stiffness of phononic unit-cell interconnections (Bergamini et al. 2014) or spatially gradient piezoelectric polarisation can be introduced to control the wave motion (Rupp et al. 2010). The microstructural composition of phononic unit-cell may also be altered by (for example) rotation of asymmetric scattering inclusions (Goffaux and Vigneron 2001) or the effective mechanical properties may be tuned by orientation of symmetric-anisotropic inclusions (Lin et al. 2011). Moreover, the thermal sensitivity of mechanical properties and density of constitutive materials is applicable in control of the phononic properties (Bayat and Gordaninejad 2016; Yao et al. 2011).

Another interesting approach is to tune the phononic bandgaps through deformation of the phononic lattice under external load, where tunability is obtained through deformation-induced gradient of both topology and stiffness (Gei et al. 2013). The strained geometry and internal stress state of the unit-cell then lead to variation of effective stiffness. The stiffness gradient can be amplified with a base material with nonlinear elastic constitutive properties. Additionally, buckling may be 'designed' into the system at certain external loads to create larger shape changes and thus considerable variation of phononic properties. Traction forces can be mechanically applied on the boundaries via physical contact, or contactless load can be applied for example through remote electromagnetic excitation of ferromagnetic inserts. Since the PhCr is supposed to experience relatively large strains for effective gradient of phononic bandgaps, appropriate constitutive properties with perfect elasticity under large deformation is desirable. Therefore, rubber-like elastomeric polymers (e.g. silicon rubber) with large deformation tolerance and almost instantaneous recoverable elastic strain are ideal candidates for such application.

Whereas the topology of PhCr unit-cell predominantly defines the bandgap properties, many research works, as discussed in preceding sections, have examined the use of topology optimisation to maximise RBW. However, very few works have considered topology optimisation of tunable bandgaps and acoustic metamaterials. The only two relevant works found in the literature are briefly discussed in the following.

Rupp et al. (2010) worked on polarisation pattern optimisation of a piezoelectric solid plate in order to design switchable wave guides or filters of vibroacoustic waves at specified frequency. Gradient-based optimisation was employed and the distribution of two contrasting material phases (with positive and negative polarisation constants leading to different acoustic impedances) throughout the domain was optimised.

Evgrafov et al. (2008) optimised the topology of tunable waveguides with the capability to bend input waves into different ports at a predefined frequency using undeformed and deformed configurations. The spatial distribution of constitutive materials Silicon-Aluminium in a square domain was identified through gradient-based optimisation. An absorbing boundary was considered for all boundaries of the design domain (typically a small part of large elastic body) and unidirectional tensile strain of 0.2% was applied for switching the wave guiding function.

However, to the best of author's knowledge, there has been no study on optimising bandgap tunability of PhCrs.

2.8 The Research Problem and Objectives

Following the presented literature, the current research is fundamentally aimed at exploring new designs for PhCrs through improved optimisation strategies and expand their application, with particular attention to the phononic lattices having finite thickness (PhPs). Multi-objective optimisation context is basically introduced to investigate the optimality of PhPs with respect to secondary objectives, other than bandgap efficiency, for enhanced functionality.

The main objectives of present research concerning design and optimisation of PhPs and relevant gaps of preceding works are discussed in the following sections.

2.8.1 Optimisation of 1D Bi-Material PhPs and Sensitivity to Unit-Cell's Filling Fraction and Aspect Ratio

Based on the literature, the optimised topology of layered 1D PhCrs has been adequately studied for bandgaps of in-plane and anti-plane bulk waves. However there is no record of optimisation of 1D PhPs for bandgaps of guided waves. Moreover, other than the work by (Dong et al. 2014) published later than this research proposal, there was no clue about the gradient of optimised topology versus filling fraction of constituents. Filling fraction of constitutive materials significantly affect the bandgap efficiency of PhCrs. As a result, maximum bandgap efficiency can be achieved for specified filling fraction to meet weight and production limitations. More importantly, a multiscale gradient phononic lattice with varying filling fraction in spatial domain can be designed to achieve maximum bandgap efficiency in micro scale and other structural objectives in macro scale. It also valuable to examine the overall bandgap efficiency of a phononic lattice including unit-cells with gradual variation of filling fractions and so bandgap properties. In the hierarchical 1D PhCr design proposed by Hussein et al. (2005) a case study was presented in which the pass bands of first 5 similar unit-cells of an optimised topology were covered by bandgaps of subsequent 5 similar unit-cells of another matched optimised topology.

Hence, the first stage of this research is concerned with:

- Topology optimisation of 1D bi-material PhPs for bandgaps of low order Lamb waves ($S_0 + A_0$).

- Studying the variation of optimised topology with respect to filling fraction of stiff inclusion through a multi-objective optimisation for different aspect ratios of PhP unit-cell.
- Examining the efficiency and multiscale functionality of a graded 1D PhP structure including various optimised topologies.

2.8.2 Optimisation of Relatively Thin and Thick Porous 2D PhPs for Structurally Worthy Bandgaps of Guided Waves

In case of porous PhPs the optimised topology for maximised RBW naturally converges to a very compliant design with maximally isolated domains having minimised connectivity. Hence, an increased topology resolution leads to enhanced bandgaps efficiency and degraded structural worthiness of optimised topology. As discussed in Sect. 2.2 managing this issue with a gradient based approach is not possible due to continuity of design domain and appearance of totally disconnected features with local resonance. In GA optimisation the solidity of topology may be controlled by the prescribed topology resolution and applying appropriate topology filter to control minimum feature sizes. However, a great technique in handling this problem is to incorporate the effective stiffness of PhP topology into optimisation algorithm in order to get designs with maximised stiffness. The feature sizes still may be additionally constrained to meet specific design manufacturing requirements.

A challenge in optimisation of porous PhCrs through GA is to efficiently dig out feasible connected topologies of design space among randomly generated candidates. Rejection of infeasible candidate topologies obviously limits the diversity of design space and overlooks those including elite patterns. Penalisation of infeasible segments of candidate topologies requires careful application of an appropriate fitness penalisation factor and very prone to compromise the optimality of solution. Filtering infeasible topologies to perfectly connected ones (if applicable) also interferes with the randomness of GA search. Consequently, development of a robust approach is essential to appropriately handle this issue. As the preceding works concerning optimisation of porous PhCrs have not detailed their approach in dealing with connectivity, it is not possible to discuss and compare them here.

Earlier studies in optimisation of porous 2D PhPs were concerned with bandgaps of asymmetric wave modes in relatively thin PhPs. Consideration of relatively thick PhPs is also worthwhile which enables making smaller features in specified thickness of background plate to control higher frequencies (shorter wavelengths). Also, complete bandgaps of guided waves and exclusive bandgaps of symmetric modes have not been addressed in any of proceeding works.

Consequently, the second stage of this research is focused on:

- Multi-objective optimisation of 2D porous PhPs for maximised RBW and maximised structural stiffness.
- Developing a robust optimisation strategy to adequately tackle connectivity issue of randomly generated porous topologies with least impact on efficiency of GA optimisation.
- Exploring relatively thin and thick optimised PhPs for maximised complete bandgaps of guided waves and exclusive bandgaps of symmetric and asymmetric modes.

2.8.3 *Optimisation of Porous PhPs for Deformation-Induced Tunability*

According to the reviewed literature in Sect. 2.7, there has been no study in relation to optimality of PhCr with tunability performance. Therefore, the third stage of present study is dedicated to design of PhP lattices with maximised bandgap tunability under specified externally imposed mechanical deformation. Bertoldi and Boyce (2008) studied the phononic bandgap tunability of such lattices under tensile strain but with infinite thickness (plain strain condition) and for specified profile of cylindrical holes.

Multi-objective optimisation of 2D porous PhPs is proposed to achieve the following simultaneous objectives when subjected to specified macromechanical equibiaxial stretch:

- Maximised RBW of guided waves in undeformed configuration.
- Maximised relative gradient of RBW (with reference to undeformed state).

Wide initial RBW and maximum tunability are typically competing objectives, motivating the use of systematic optimisation. The relative gradient of RBW could be maximised in descending or ascending fashion. The former introduces degraded (ideally closed) bandgap after deformation and the latter potentially results in widened bandgap by deformation.

2.9 Thesis Structure

Following the aforementioned research scope and objectives in Chap. 1, the thesis is structured as follows:

In Chap. 3 the implemented optimisation framework is described. Optimisation algorithms and their characteristics are explained and compared, and relevance of GA optimisation to proposed research is argued. Then the topology mapping and definition of binary design variables for desired 1D and 2D PhPs are presented. The constitutive formulations of PhP unit-cell are basically explained. Modal band analysis of unit-cell is particularly described for calculation of bandgap objective function. Detailed discussion regarding evaluation of secondary objectives is left for relevant chapters.

In Chap. 4, the optimised topology of 1D bi-material PhPs to attain maximum relative bandgap width of low order Lamb waves is computationally investigated. The evolution of optimised topology with respect to filling fraction of constituents, alternatively stiff scattering inclusion, is explored. Multi-objective genetic algorithm (GA) is employed in this research in conjunction with finite element method (FEM) for topology optimisation of silicon-tungsten PhCr plate unit-cells. A specialised FEM model is developed and verified for dispersion analysis of guided waves and calculation of modal response. Being inspired by the obtained optimum topologies, definite and easy to produce topologies are proposed with enhanced bandgap efficiency as compared to regular centric unit-cells. Finally a few cases are introduced to evaluate the frequency response of finite PhCr plate structures produced by achieved topologies and also to confirm the reliability of calculated modal band structures. Cases made by consecutive unit-cells of different filling fraction are examined in order to attest the bandgap efficiency and multiscale functionality of such graded PhCr plate structures.

Topology optimisation of 2D porous PhPs is presented in Chap. 5. In order to study the bandgap efficiency of PhP unit-cell with respect to its structural worthiness, the in-plane stiffness is incorporated in optimisation algorithm as an opposing objective to be maximised. Thick and relatively thin Polysilicon PhP unit-cells with square symmetry are studied. Non-dominated sorting genetic algorithm NSGA-II is employed for this multi-objective optimisation problem and modal band analysis of individual topologies is performed through finite element method. Specialised topology initiation, evaluation and filtering are applied to achieve refined feasible topologies without penalising the randomness of genetic algorithm (GA) and diversity of search space. Selected Pareto topologies are presented and gradient of RBW and elastic properties in between the two Pareto front extremes are investigated. Finally, steady state and transient frequency response of finite thin PhP structures of selected Pareto topologies are studied and validity of obtained bandgaps is confirmed.

In Chap. 6 a topology refinement study is performed by doubling the topology resolution. The optimisation algorithm introduced in Chap. 5 is also improved for higher computational efficiency and controlled evolution of topologies in last stages of optimisation. The coarse topology is first obtained through improved algorithm and afterwards is refined in a secondary stage. The Pareto fronts are presented and analysed and relative performance of achieved topologies is evaluated to explore the efficiency of topology refinement. Various aspect ratios are optimised and the distinction and bandgap-stiffness efficiency of topologies (for the same unit-cell width and different aspect ratios) are investigated.

In Chap. 7 porous 2D PhPs with optimised bandgap efficiency and tunability of guided waves are computationally explored and evaluated. Various tunability targets are defined to enhance or degrade complete bandgaps of guided waves through macroscopic tensile deformation. Elastomeric hyperelastic material is considered which enables recoverable micromechanical deformation under tuning finite stretch. Phononic tunability through stable deformation of phononic lattice is specifically required and so any topology showing buckling instability under assumed deformation is disregarded. NSGA-II is adopted for evolutionary multi-objective topology optimisation of hypothesized tunable PhP with square symmetric unit-cell and relevant topologies are analysed through finite element method. Following earlier studies by the authors, specialised GA algorithm, topology mapping, assessment and analysis techniques are employed to get feasible porous topologies of assumed thick PhP, efficiently.

Then Chap. 8 is dedicated to experimental evaluation of selected optimised PhPs. Various topologies with different relative bandgap efficiency and structural stiffness are selected and evaluated. The transmission spectrum of frequency sweep and random white noise excitations through produced PhP lattices are measured and compared with calculated modal band structures. Furthermore, tensile tests are carried out and the equivalent elastic properties and microstructural deformation of selected topologies are measured by means of extensometers and also through stereovision digital image correlation (3D-DIC). The measured elastic properties are evaluated in comparison with computed homogenised properties.

Finally in Chap. 9 conclusions are made and suggestions for future work are marked out.

2.10 Concluding Remarks

The existing literature concerning topology optimisation of AMMs was categorised and described based on various optimisation objectives considered for achieving tailored vibroacoustic behaviour. Studies introducing prescribed shapes and topologies for PhCrs lattices were addressed. Moreover, topology optimisation studies in relation to PhCrs were detailed and the lack of literature was discussed.

Consequently, the research objective was defined as exploring new designs for PhCrs through improved optimisation strategies and expand their application, with particular attention to the phononic lattices having finite thickness (PhPs). Multi-objective optimisation context is introduced to investigate the optimality of PhPs with respect to secondary objectives, other than bandgap efficiency, for enhanced functionality. The research problems were specified and discussed.

Although the key contributions of the literature within the scope of this thesis were adequately reviewed in this chapter, other relevant works are discussed in further chapters were necessary.

References

Bayat A., & Gordaninejad, F. (2016) A thermally tunable phononic crystal. In *SPIE Smart Structures and Materials + Nondestructive Evaluation and Health Monitoring* (pp. 98000 W–98000 W). International Society for Optics and Photonics.

Bergamini, A., Delpero, T., Simoni, L. D., Lillo, L. D., Ruzzene, M., & Ermanni, P. (2014). Phononic crystal with adaptive connectivity. *Advanced Materials, 26*(9), 1343–1347.

Bertoldi, K., & Boyce, M. C. (2008). Wave propagation and instabilities in monolithic and periodically structured elastomeric materials undergoing large deformations. *Physical Review B, 78*(18), 184107.

Bilal, O. R., & Hussein, M. I. (2011). Ultrawide phononic band gap for combined in-plane and out-of-plane waves. *Physical Review E, 84*(6), 065701.

Bilal, O. R., & Hussein, M. I. (2012). Topologically evolved phononic material: Breaking the world record in band gap size. In *Photonic and Phononic Properties of Engineered Nanostructures* (pp. 826911–826917). International Society for Optics and Photonics.

Charles, C., Bonello, B., & Ganot, F. (2006, December 22). Propagation of guided elastic waves in 2D phononic crystals. In *Ultrasonics,* (Vol. 44, Supplement, no. 0, pp. e1209–e1213).

Chen, S., Gonella, S., Chen, W., & Liu, W. K. (2010). A level set approach for optimal design of smart energy harvesters. *Computer Methods in Applied Mechanics and Engineering, 199*(37–40), 2532–2543.

Chen, J.-J., & Han, X. (2010). The propagation of Lamb waves in one-dimensional phononic crystal plates bordered with symmetric uniform layers. *Physics Letters A, 374*(31–32), 3243–3246.

Dong, H.-W., Su, X.-X., & Wang, Y.-S. (2014a). Multi-objective optimization of two-dimensional porous phononic crystals. *Journal of Physics. D. Applied Physics, 47*(15), 155301.

Dong, H.-W., Su, X.-X., Wang, Y.-S., & Zhang, C. (2014b). Topological optimization of two-dimensional phononic crystals based on the finite element method and genetic algorithm. *Structural and Multidisciplinary Optimization, 50*(4), 593–604.

Dong, H.-W., Su, X.-X., Wang, Y.-S., & Zhang, C. (2014c). Topology optimization of two-dimensional asymmetrical phononic crystals. *Physics Letters A, 378*(4), 434–441.

Du, J., & Olhoff, N. (2010). Topological design of vibrating structures with respect to optimum sound pressure characteristics in a surrounding acoustic medium. *Structural and Multidisciplinary Optimization, 42*(1), 43–54.

El-Sabbagh, A., Akl, W., & Baz, A. (2008). Topology optimization of periodic Mindlin plates. *Finite Elements in Analysis and Design, 44*(8), 439–449.

Evgrafov, A., Rupp, C. J., Dunn, M. L., & Maute, K. (2008). Optimal synthesis of tunable elastic wave-guides. *Computer Methods in Applied Mechanics and Engineering, 198*(2), 292–301.

Fan Li, Y., Huang, X., Meng, F., & Zhou, S. (2016). Evolutionary topological design for phononic band gap crystals. *Structural and Multidisciplinary Optimization*, 1–23.

Gazonas, G. A., Weile, D. S., Wildman, R., & Mohan, A. (2006). Genetic algorithm optimization of phononic bandgap structures. *International Journal of Solids and Structures, 43*(18–19), 5851–5866.

Gei, M., Bigoni, D., Movchan, A., & Bacca, M. (2013). Band-gap properties of prestressed structures. In *Acoustic metamaterials*(pp. 61–82). Berlin: Springer.

Goffaux, C., & Vigneron, J. P. (2001). Theoretical study of a tunable phononic band gap system. *Physical Review B, 64*(7), 075118.

Halkjær, S., Sigmund, O., & Jensen, J. S. (2005). Inverse design of phononic crystals by topology optimization. *Zeitschrift für Kristallographie, 220*(9–10), 895–905.

Halkjær, S., Sigmund, O., & Jensen, J. S. (2006). Maximizing band gaps in plate structures. *Structural and Multidisciplinary Optimization, 32*(4), 263–275.

Hu, X., Shen, Y., Liu, X., Fu, R., & Zi, J. (2004). Superlensing effect in liquid surface waves. *Physical Review E, 69*(3), 030201.

Huang, Y., Liu, S., & Zhao, J. (2013). Optimal design of two-dimensional band-gap materials for uni-directional wave propagation. *Structural and Multidisciplinary Optimization, 48*(3), 487–499.

Huang, P., Yao, Y., Wu, F., & Zhang, X. (2014). Numerical study of Lamb waves band structure in one-dimensional phononic crystal slabs with the anti-symmetric boundary structure. *Journal of Applied Physics, 115*(6), 063510.

Hussein, M. I., Hamza, K., Hulbert, G. M., & Saitou, K. (2007). Optimal synthesis of 2D phononic crystals for broadband frequency isolation. *Waves in Random and Complex Media, 17*(4), 491–510.

Hussein, M. I., Hamza, K., Hulbert, G. M., Scott, R. A., & Saitou, K. (2006). Multiobjective evolutionary optimization of periodic layered materials for desired wave dispersion characteristics. *Structural and Multidisciplinary Optimization, 31*(1), 60–75.

Hussein, M. I., Hulbert, G. M., & Scott, R. A. (2002). Tailoring of wave propagation characteristics in periodic structures with multilayer unit cells. In *Proceedings of the 7th American Society of Composites Technical Conference,* [CD ROM], West Lafayette, Indiana.

Hussein, M. I., Hulbert, G. M., & Scott, R. A. (2005). Hierarchical design of phononic materials and structures. In *Proceedings of 2005 ASME International Mechanical Engineering Congress and R&D Expo* (pp. 1–10).

Kao, C., Osher, S., & Yablonovitch, E. (2005). Maximizing band gaps in two-dimensional photonic crystals by using level set methods. *Applied Physics B, 81*(2–3), 235–244.

Khelif, A., Aoubiza, B., Mohammadi, S., Adibi, A., & Laude, V. (2006). Complete band gaps in two-dimensional phononic crystal slabs. *Physical Review E, 74*(4), 046610.

Larsen, A., Laksafoss, B., Jensen, J., & Sigmund, O. (2009). Topological material layout in plates for vibration suppression and wave propagation control. *Structural and Multidisciplinary Optimization, 37*(6), 585–594.

Li, J., Wang, Y.-S. & Zhang, C. (2009). Finite element method for analysis of band structures of phononic crystal slabs with archimedean-like tilings. In *IEEE International Ultrasonics Symposium* (pp. 1548–1551).

Li, D., Zigoneanu, L., Popa, B.-I., & Cummer, S. A. (2012). Design of an acoustic metamaterial lens using genetic algorithms. *The Journal of the Acoustical Society of America, 132*(4), 2823–2833.

Lin, C.-M., Hsu, J.-C., Senesky, D. G., & Pisano, A. P. (2014) Anchor loss reduction in ALN Lamb wave resonators using phononic crystal strip tethers. In *Frequency Control Symposium (FCS), 2014 IEEE International* (pp. 1–5).

Lin, S.-C. S., & Huang, T. J. (2011). Tunable phononic crystals with anisotropic inclusions. *Physical Review B, 83*(17), 174303.

Liu, Z.-F., Wu, B., & He, C.-F. (2014). Band-gap optimization of two-dimensional phononic crystals based on genetic algorithm and FPWE. *Waves in Random and Complex Media*, no. ahead-of-print, pp. 1–20.

Lu, L., Yamamoto, T., Otomori, M., Yamada, T., Izui, K., & Nishiwaki, S. (2013). Topology optimization of an acoustic metamaterial with negative bulk modulus using local resonance. *Finite Elements in Analysis and Design, 72,* 1–12.

Manktelow, K. L., Leamy, M. J., & Ruzzene, M. (2013). Topology design and optimization of nonlinear periodic materials. *Journal of the Mechanics and Physics of Solids, 61*(12), 2433–2453.

Matar, O. B., Vasseur, J., & Deymier, P. A. (2013). Tunable phononic crystals and metamaterials. In *Acoustic metamaterials and phononic crystals* (pp. 253–280). Berlin: Springer.

Men, H., Freund, R. M., Nguyen, N. C., Saa-Seoane, J., & Peraire, J. (2013). Designing phononic crystals with convex optimization. In *ASME 2013 International Mechanical Engineering Congress and Exposition* (pp. V014T015A047–V014T015A047). American Society of Mechanical Engineers.

Meng, F., Huang, X., & Jia, B. (2015). Bi-directional evolutionary optimization for photonic band gap structures. *Journal of Computational Physics, 302,* 393–404.

Miniaci, M., Marzani, A., Testoni, N., & De Marchi, L. (2015). Complete band gaps in a polyvinyl chloride (PVC) phononic plate with cross-like holes: Numerical design and experimental verification. *Ultrasonics, 56,* 251–259.

Mohammadi, S. (2010). Phononic band gap micro/nano-mechanical structures for communications and sensing applications. Georgia Institute of Technology.

Niu, B., Yan, J., & Cheng, G. (2009). Optimum structure with homogeneous optimum cellular material for maximum fundamental frequency. *Structural and Multidisciplinary Optimization, 39*(2), 115–132.

Olhoff, N., Niu, B., & Cheng, G. (2012). Optimum design of band-gap beam structures. *International Journal of Solids and Structures, 49*(22), 3158–3169.

Olsson Iii, R. H., & El-Kady, I. F. (2009). Microfabricated phononic crystal devices and applications. *Measurement Science & Technology, 20*(1), 012002.

Park, J. H., Ma, P. S., & Kim, Y. Y. (2015). Design of phononic crystals for self-collimation of elastic waves using topology optimization method. *Structural and Multidisciplinary Optimization, 51*(6), 1199–1209.

Pennec, Y., Vasseur, J. O., Djafari-Rouhani, B., Dobrzyński, L., & Deymier, P. A. (2010). Two-dimensional phononic crystals: Examples and applications. *Surface Science Reports, 65*(8), 229–291.

Rupp, C. J., Dunn, M. L., & Maute, K. (2010). Switchable phononic wave filtering, guiding, harvesting, and actuating in polarization-patterned piezoelectric solids. *Applied Physics Letters, 96*(11), 111902.

Rupp, C. J., Evgrafov, A., Maute, K., & Dunn, M. (2007). Design of phononic materials/structures for surface wave devices using topology optimization. *Structural and Multidisciplinary Optimization, 34*(2), 111–121.

Shu, L., Wang, M. Y., Fang, Z., Ma, Z., & Wei, P. (2011). Level set based structural topology optimization for minimizing frequency response. *Journal of Sound and Vibration, 330*(24), 5820–5834.

Sigmund, O., & Jensen, J. S. (2003). Systematic design of phononic band-gap materials and structures by topology optimization. *Philosophical Transactions of the Royal Society of London A, 361,* 1001–1019.

Su, Z., & Ye, L. (2009). *Identification of damage using lamb waves: From fundamentals to applications*. Berlin: Springer.

Vatanabea, S. L., & Silva, E. C. N. (2011). Design of phononic band gaps in functionally graded piezocomposite materials by using topology optimization. *Behavior and Mechanics of Multifunctional Materials and Composites*, pp. 797811-1–797811-10.

Veidt, M., Ng, C.-T., Hames, S., & Wattinger, T. (2008). Imaging laminar damage in plates using Lamb wave beamforming. *Advanced Materials Research, 47,* 666–669.

Wang, X., Chen, S., & Zuo, L. (2015). On design of mechanical metamaterials using level-set based topology optimization. In *ASME 2015 International Design Engineering Technical Conferences and Computers and Information in Engineering Conference*.

Wang, Y., Gao, J., Luo, Z., Brown, T & Zhang, N. (2016a). Level-set topology optimization for multimaterial and multifunctional mechanical metamaterials. *Engineering Optimization*, pp. 1–21.

Wang, Y., Luo, Z., Zhang, N., & Kang, Z. (2014). Topological shape optimization of microstructural metamaterials using a level set method. *Computational Materials Science, 87,* 178–186.

Wang, Y., Luo, Z., Zhang, N., & Wu, T. (2016b). Topological design for mechanical metamaterials using a multiphase level set method. *Structural and Multidisciplinary Optimization*, pp. 1–16.

Wu, T.-T., Hsu, J.-C., & Sun, J.-H. (2011). Phononic plate waves. *IEEE Transactions on Ultrasonics, Ferroelectrics and Frequency Control, 58*(10), 2146–2161.

Yang, X. W., Lee, J. S., & Kim, Y. Y. (2016). Effective mass density based topology optimization of locally resonant acoustic metamaterials for bandgap maximization. *Journal of Sound and Vibration*.

Yao, Y., Wu, F., Zhang, X., & Hou, Z. (2011). Thermal tuning of Lamb wave band structure in a two-dimensional phononic crystal plate. *Journal of Applied Physics, 110*(12), 123503.

Yilmaz, C., Hulbert, G. M., & Kikuchi, N. (2007). Phononic band gaps induced by inertial amplification in periodic media. *Physical Review, 76*(5), 054301–054309.

Yu, K., Chen, T., & Wang, X. (2013). Large band gaps in two-dimensional phononic crystals with neck structures. *Journal of Applied Physics, 113*(13), 134901.

Chapter 3
Optimisation Framework and Fundamental Formulation

3.1 Introduction

The implemented optimisation framework is first discussed in this chapter. Optimisation algorithms and their characteristics are explained and compared. The advantages of GA for optimisation of PhCrs and in particular for proposed research problem are argued. Then the topology mapping and definition of binary design variables for desired 1D and 2D PhPs are presented.

Since modal band analysis is a common objective of all of this research stages, its FEM calculation is further discussed in this chapter. Constitutive stress-strain formulation of the elastic base material and Navier's equation of equilibrium are given first. Then the FEM formulation and relevant periodic boundary conditions for modal band analysis of the PhP unit-cell are discussed. Detailed formulation and also evaluation procedure of other objectives will be discussed later on, in relevant chapters.

3.2 Optimisation Methods and Genetic Algorithm

Optimisation methods are classified into two major groups: Deterministic and Stochastic algorithms (Cavazzuti 2013).

Deterministic (i.e. gradient based) optimisation is based on mathematical programming and finds the optimised solution by taking into account the gradient of objective function. Sensitivity analysis is performed in the vicinity of a sample point to move towards optimum solution. Hence, a reliable objective function of continuous design variables is required to perform optimisation. An objective function is normally estimated explicitly based on sensitivity information of current solution and the estimated sub problem is analytically optimised and a local

S. Hedayatrasa, *Design Optimisation and Validation of Phononic Crystal Plates for Manipulation of Elastodynamic Guided Waves*, Springer Theses,
https://doi.org/10.1007/978-3-319-72959-6_3

optimum is estimated to provide the initial guess of next iteration (Svanberg 1987). The iteration repeats until the problem is converged.

Bi-directional evolutionary structural optimisation (BESO) is also an alternative gradient based approach which evolves material distribution based on the usefulness (optimality criteria or sensitivity information) (Huang and Xie 2010; Radman 2013). BESO has been reliably used for topology optimisation of phononic and photonic crystals (Fan Li et al. 2016; Meng et al. 2015).

Level-set method is another alternative approach for gradient based topology optimisation (Wang et al. 2003). The interface of different material phases is initially prescribed by a level set function which dynamically changes at different points to form an optimised shape based on a speed vector correlated to the gradient of objective function. As presented in Chap. 2, the level set method has been efficiently implemented and developed for topology optimisation of metamaterials (Kao et al. 2005; Wang et al. 2016a, b).

The main advantage of gradient based optimisation is its faster convergence and lower computational cost compared with the stochastic optimisation. However because of its gradient based nature may get stuck in a local optimum and several initial guesses may be needed to make sure about its reliability.

In contrary, stochastic optimisation algorithms are random in nature with no need to derivative information. These algorithms generally work on a population of initial sample points and evolve them towards optimum solution by mimicking natural methods. Many stochastic optimisation techniques have been introduced and examined in the literature such as genetic algorithm (GA), simulated annealing, Tabu search and ant colony optimisation. Among stochastic techniques GA has been widely and reliably employed for topology optimisation of structures, multi-objective problems and design of PhCrs in particular.

GA is an evolutionary stochastic algorithm that mimics the metaphor of natural-biological evolution. GA operates on a population of potential solutions applying the principle of survival of the fittest to possibly produce better and better approximations to a solution. At each generation, a new set of approximations is created by the process of selecting individuals according to their level of fitness in the problem domain and breeding them together using operators borrowed from natural genetics. This process leads to the evolution of populations of individuals that are better suited to their environment than the individuals that they were created from, just as in natural adaptation (Rennera and Ekart 2003). Crossover operation explores design space for better individuals by retaining good parts of mating parents. Mutation, on the other hand, exploits the design space by maintaining the diversity of population and reduces the likelihood of premature convergence.

The main advantages of GA as a stochastic optimisation compared to deterministic optimisation are as follows:

- GA works with objective function or fitness information only and do not rely on derivative information. So it is less mathematically complicated and also it is capable to evaluate discontinuous functions.

- GA is an implicitly parallel algorithm which works on a group of design points instead of a single point and searches many areas of the design space at once. So it is less likely to stick with a local optimum.
- GA relies on a set of probabilistic rules instead of deterministic ones, allowing for some of the decisions during the search to be made by drawing random numbers and avoid local optimal points easier.
- GA can easily be implemented for multi-objective optimisation, while multi-objective implementation of deterministic optimisation is not efficient.

These characteristics have made GA a desirable strategy for intelligent design of discrete and strongly multimodal problems, with the possibility of suboptimal solutions. Computational intensity and propensity to converge prematurely are main shortcomings associated with GA which may be avoided by utilising appropriate strategies. Moreover, the physics behind the problem, its multi-objective definition and complexity may make it essentially necessary to implement a stochastic optimisation algorithm like GA.

3.3 Genetic Algorithm Topology Optimisation of PhPs

As explained in Chap. 2, both gradient based and stochastic GA optimisation techniques have been used in the literature for topology optimisation of PhCrs. However, topology optimisation of PhCrs, particularly within specified scope of this research, has its specific characteristics and limitations. Therefore evolutionary based optimisation through GA is employed for topology optimisation of PhP unit-cell, rather than gradient based approach.

Principally, gradient based topology optimisation on a continuous design space of design variables, i.e. material properties, provides the best material layout to achieve the desired objective. The optimality may be fulfilled by pseudo domains with intermediate artificial material properties. Such intermediate domains may be achieved by considering a multiscale design with equivalent homogenised properties, or may be faded via appropriate penalisation approach. But, in general, the optimum topology of PhCrs providing widest bandgap naturally converges to distinct material layout as highest material contrast is favoured for maximisation of interfacial reflections. So, the discrete nature of GA suits the desirable design space of current optimisation problem. However, GA optimisation is implemented in this work mainly due to the following critical requirements and restrictions:

- When total bandgap within a number of modal branches is to be maximised (instead of a single gap) then GA is a robust approach which handles the multimodality of such problem having disjoint design space (Sigmund 2011).
- Available gradient based approaches require the continuity of design domain (unit-cell) to be maintained during optimisation to avoid singularities and also to calculate the sensitivities over the entire domain. Hence, for optimisation of

porous phononic crystals a compliant material should be considered as the void area. This leads to appearance of artificially optimised topologies including isolated solid domains (islands) which are impractical. This is mainly because local resonance and wave scattering enabled by isolated solid features (with weak connection to the surroundings through non-existing compliant material) artificially produce wide bandgaps at low frequencies (lower RBW) as desired. This issue has not been observed in common structural topology optimisation problems for e.g. minimised compliance at specified filling fraction. Actually in such problems any disconnected solid feature which has no contribution in objective (and any weakly connected feature with minor contribution) is naturally eliminated.

- This research is mainly concerned with maximising RBW of phononic crystals by taking into account the effect of filling fraction, stiffness and tunability. Therefore the optimisation problem has multi-objective nature for which GA is a great choice. A non- dominated population of optimised topologies are obtained in a single run which define the gradient of optimised solution with respect to objective function values.

And some other aspects making GA an appropriate approach are:

- In GA the topology of porous unit-cell is defined with a binary string in which any cell indicates the solid or void status of a topology pixel. Hence, the morphology and connectivity of topology can be evaluated. Then due to the random nature of GA, specific operations can be employed to mutate topologies towards connected ones with desired features.
- Because of the discreteness of GA, the void section of porous unit-cell can be eliminated so that only the solid section of unit-cell is analysed for fitness evaluation purpose. This can significantly reduce the computational cost. Of course, efficient implementation of this technique depends on capabilities of fitness evaluation approach and analysis tool.

The application of GA for optimisation of PhCrs is adequately described in Gazonas et al. (2006). Some other works have also taken the advantage of GA for optimisation of 1D and 2D bi-material phononic bandgaps of longitudinal, flexural and bulk waves (Bilal and Hussein 2011; Gazonas et al. 2006; Hussein et al. 2006, 2007).

3.3.1 Implemented Multi-objective Optimisation Algorithm (NSGA-II)

Definition of multiple objectives in optimisation problem principally leads to non-uniqueness of the solution. Instead of a single solution a set of solutions exist, widely known as Pareto optimal solutions. Pareto solutions are located on the most dominant frontier of all alternate solutions in the solution space (Pareto front). The objectives magnitude varies over the Pareto front; however, none of the Pareto solutions is dominant to the others and the best Pareto solution depends on the design requirements and preferences.

A multi-objective problem may be tackled by incorporating all objectives into a single combined objective. However, the efficiency of this approach depends on the nature of problem and obtained solution is mainly affected by the dominancy (weight) of different objectives. Moreover, a single solution is provided and gradient of optimal solution versus objectives may be studied by running several optimisations with different objective combinations and weighting factors.

Evolutionary GA multi-objective optimisation algorithms evolve a population of alternate solutions by taking into account all objectives separately, and finally give a set of Pareto optimised solutions as the Pareto front in a single run. Various multi-objective GA algorithms have been introduced in the literature. Modified non-dominated sorting GA (NSGA-II) is one of the best algorithms introduced by (Pratap et al. 2002) which is employed in current study for multi-objective optimisation of PhPs.

NSGA-II improved the original NSGA algorithm by reducing its computational cost, incorporating elitism and ensuring the diversity of Pareto front. NSGA-II sorts GA population based on non-dominated fitness and crowding distance. In this way the designs with the most dominated fitness are placed at the first rank (Pareto front) and among them those with highest crowding distance (in less crowded region) are preferred.

For domination sorting of population in NSGA-II, any solution is searched for two entities: (i) the dominate count as the number of solutions which dominate the solution, and (ii) the dominated solution set as the group of solutions which the solution dominates. A solution dominates another one provided that it has better fitness for both objectives. If exceptionally both solutions have the same function values for one objective, then the solution with higher value for the other objective is dominant. Obviously the solutions with zero dominate count locate in the first rank (Pareto front). In the next step, the dominate count of the dominated solution set of first ranked solutions is reduced by one in which the new zero dominate counts define the second rank. This procedure continues until all front levels are defined.

The crowding-distance on the other hand is a spread measure of NSGA-II to enforce even distribution and diversity of optimised solutions within each front level. In order to compute the crowding-distance, the population needs to be sorted according to the objective function value in ascending order of magnitude. Then for each objective function, the extreme solutions with highest and lowest fitness value are assigned an infinite crowding-distance to ensure they are not degraded by the crowdedness criteria. The crowding-distance of intermediate solutions is defined as the absolute normalised difference of the objective values of its two adjacent solutions. Then the overall crowding-distance is calculated by summation of distance values for all objectives.

The basic steps of NSGA-II multi-objective optimisation algorithm implemented in this research are as follows:

1. Create a random initial population. Any individual (chromosome) is a bit-string vector (containing zeros and ones) which its cells (genes) define the material type (design variable) over the discretised unit-cell (design domain) of the assumed bi-material topology.
2. Check feasibility of topology introduced by any individual in the current population, if necessary, and define the objective functions (fitness) of feasible individuals.
3. Define the rank and crowding distance of individuals and sort them accordingly, as discussed earlier in this section.
4. Create parent pool through tournament selection.
5. Two randomly selected individuals are compared and the one with higher rank is selected as candidate parent. If both have the same rank, the one with higher crowding distance is selected.
6. Create offspring population by applying single point crossover with certain probability on couples of randomly selected parents providing two Children per couple.
7. A random probability value is first generated for selected couple and if its magnitude is equal or bigger than specified crossover probability, then the crossover operation is performed. A random crossover point (cell number) is selected at which the vectors of mating couple are divided and swapped.
8. Mutate offspring population with specified mutation probability.
9. A random probability vector is generated for each individual topology. Those cells of topology having probability equal or more than specified mutation probability are inversed (i.e. 0–1 and vice versa).
10. If necessary, check the morphology of offspring population and apply objective mutations randomly in order to obtain feasible topologies with desirable features.
11. Produce intermediate population by accumulating offspring and current population and sorting them based on ranking and crowding distance. This step ensures the elitism of optimised solutions by maintaining the elite ones.
12. Select topologies of next generation based on their ranking, and if the first rank is bigger than required population then select based on crowding distance.
13. Go to Step-2 if termination condition (prescribed number of generations) is not satisfied, otherwise stop.

More detailed algorithms are delivered according to the specific operations needed for different optimisation steps in the relevant chapters. The above optimisation algorithm is coded in MATLAB which calls an internal or external code to evaluate the fitness of candidate topologies. Finite element method (FEM) is employed for calculation of objective functions. A FEM code is developed in MATLAB for analysis of 1D bi-material PhPs, and a macro file is coded in ANSYS APDL FEM solver for convenient and efficient analysis of 2D PhPs.

3.3.2 Topology Mapping and Design Variables

In this section discretisation and mapping of PhP topologies are discussed so that any candidate topology is characterised by a binary vector of design variables to be used during optimisation process. The vector assigns the material properties of topology cells based on assumed unit-cell shape and symmetry.

Bi-material layered 1D PhP unit-cell with length a and thickness h is evenly discretised along the length and a binary design variable is assigned to each layer as shown in Fig. 3.1a. So the relevant binary vector $\mathbf{\Phi}$ identifying the unit-cell topology is defined in which 1 stands for stiff base material and 0 stands for soft base material, as given in Fig. 3.1a for this arbitrary topology. If symmetric topology like the one shown in Fig. 3.1b is assumed, then the topology vector is reduced by half.

For porous 2D PhPs, unit-cell with square symmetry is basically assumed and uniformly discretised into square cells as schematically presented in Fig. 3.2 for an arbitrary porous topology. Due to prescribed square symmetry, the design domain is limited to the irreducible triangular shaded area with 55 independent design variables as numbered in Fig. 3.2b, and coded with binary design variables in Fig. 3.2d. Figure 3.2b shows how design variables are assigned to each cell of the arbitrary unit-cell topology in which 1 stands for solid background and 0 stands for void area of porous PhP.

However, the design variable domain should not necessarily have the same spatial distribution as the unit-cell topology cells. The design variable may be located anywhere like on topology cell centroids or nodes and mapped to the design

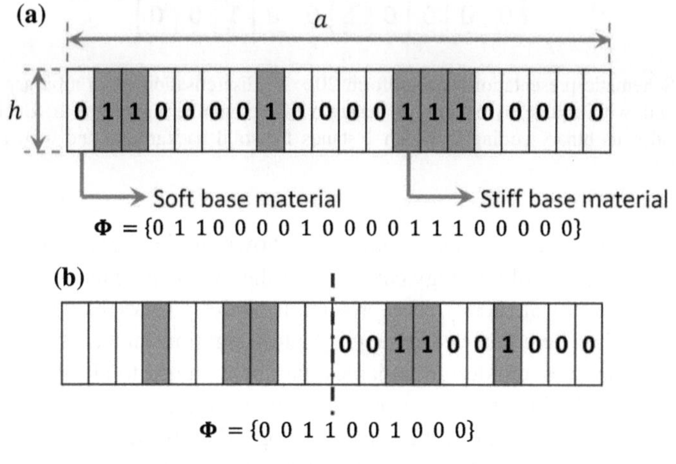

Fig. 3.1 Schematic presentation of discretisation of an arbitrary 1D bi-material PhP unit-cell into 20 layers and relevant binary vectors $\mathbf{\Phi}$, **a** No symmetry, **b** symmetric

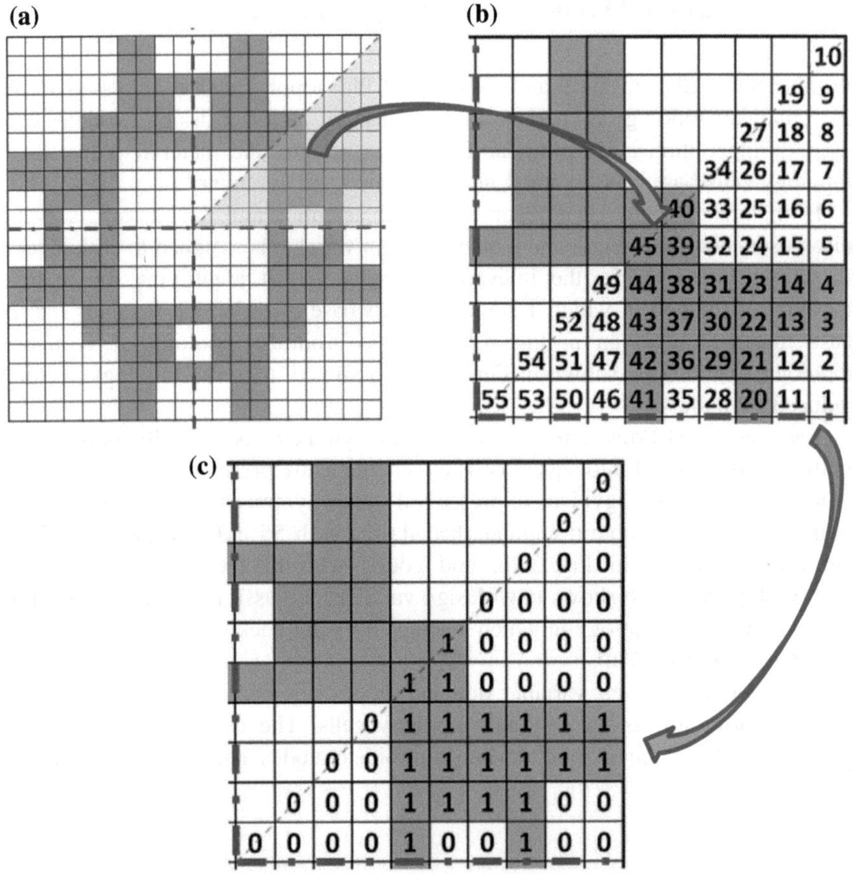

Fig. 3.2 Schematic presentation of **a** uniform 20×20 discretisation of an arbitrary porous 2D PhP unit-cell with square symmetry, **b** design variable numbering of the irreducible design domain, and **c** its binary coding in which 1 stands for solid background and 0 stands for void (porous) area

domain with specified projection radius r_p as shown in Fig. 3.3a, b. In this way the minimum feature size of topology can be controlled while maintaining the topology resolution (Guest et al. 2004). The projection of design variables to surrounding elements can then be controlled by appropriate weighting functions dependents on the radius. Moreover, the design variables may be assigned to all or every few cell nodes or centroids. Figure 3.3c, d present how a reduced design variable size on every other node may lead to the same topology due to the redundancy of intermediate design variables. Moreover, the capability to introduce smaller features (relative to the projection diameter) is shown in Fig. 3.3d by incorporating and altering marginal design variables. The redundancy of design variables caused by

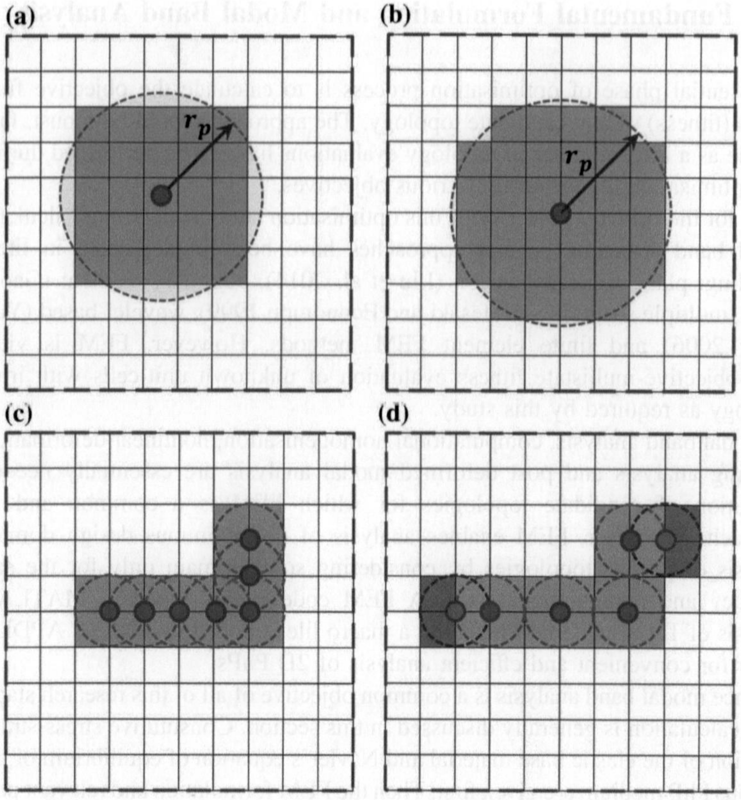

Fig. 3.3 Schematic presentation of mapping binary design variable over a discretised area of unit-cell (Guest and Smith Genut 2010); design variable assigned to **a** cell centroid and **b** cell corners with projection radius r_p; **c** nodal design variables assigned to all nodes; **d** nodal design variables assigned to every other node and introducing small features by adding-altering marginal design variables

overlapped projection areas may be handled in order to decrease the design space and so enhance the efficiency of optimisation process (Guest and Smith Genut 2010) which is out of scope of this study.

In this study, primarily regular design variables on all topology centroids are considered to be mapped on just one cell (i.e. projection radius of half cell size) as shown in Fig. 3.2. Then the topologies are redefined with appropriate nodal design variables to perform topology refinement. Further detailed discussions are given in relevant chapters.

It is noteworthy that due to computational intensity and huge number of fitness evaluations needed for proposed multi-objective optimisation of PhPs, considering very fine meshes and so large projection radiuses during optimisation is very costly and impractical.

3.4 Fundamental Formulation and Modal Band Analysis

An essential phase of optimisation process is to calculate the objective function values (fitness) of any candidate topology. The approach should be robust, fast and reliable as a large number of topology evaluations have to be performed during the GA optimisation algorithm for various objectives.

As for the primary objective of this optimisation study concerning calculation of modal band structure, several approaches have been implemented in literature including: plane wave expansion (Liu et al. 2014), boundary element (Gao et al. 2013), multiple scattering (Kafesaki and Economou 1999), wavelet-based (Yan and Wang 2006) and finite element FEM methods. However, FEM is vital for multi-objective-multistate fitness evaluation of unknown unit-cells with irregular topology as required by this study.

Modal band analysis, computational homogenisation, nonlinear deformation and buckling analysis and post deformed modal analysis are essentially needed for evaluation of candidate topologies for which FEM is a common and robust approach. Moreover, FEM enables analysis of discontinuous design domain for analysis of porous topologies by considering solid domain only for the sake of accuracy and computational cost. A FEM code is developed in MATLAB for analysis of 1D bi-material PhPs, and a macro file is coded in ANSYS APDL FEM solver for convenient and efficient analysis of 2D PhPs.

Since modal band analysis is a common objective of all of this research stages, its FEM calculation is generally discussed in this section. Constitutive stress-strain formulation of the elastic base material and Navier's equation of equilibrium of heterogeneous PhP medium are given first. Then the FEM formulation and relevant periodic boundary conditions for modal band analysis of the PhP unit-cell are discussed.

3.4.1 Elastic Constants and Stress-Strain Relations (Tsai 1992)

The linear stress-strain constitutive relation for small deformation of a linearly elastic material in tensorial notation is:

$$\sigma_{ij} = C_{ijkl}\varepsilon_{kl} \tag{3.1}$$

where the indexes can take any value in $\{1, 2, 3\}$, C_{ijkl} is the matrix of elastic constants, σ_{ij} is stress tensor and ε_{kl} is the infinitesimal strain tensor for small displacement vector $\mathbf{u}$:

$$\varepsilon_{kl} = \frac{1}{2}\left(u_{k,l} + u_{l,k}\right) \tag{3.2}$$

The comma (,) in the index means derivative to the relevant coordination direction e.g. $u_{i,j} = \frac{\partial u_i}{\partial x_j}$. Due to the symmetry of stress and strain tensors, the matrix of elastic constants C_{ijkl} is symmetric. So the equation can be rewritten in matrix form as follows:

$$\begin{Bmatrix} \sigma_{11} \\ \sigma_{22} \\ \sigma_{33} \\ \sigma_{23} \\ \sigma_{31} \\ \sigma_{12} \end{Bmatrix} = \begin{bmatrix} C_{11} & C_{12} & C_{13} & C_{14} & C_{15} & C_{16} \\ & C_{22} & C_{23} & C_{24} & C_{25} & C_{26} \\ & & C_{33} & C_{34} & C_{35} & C_{36} \\ & & & C_{44} & C_{45} & C_{46} \\ & Sym & & & C_{55} & C_{56} \\ & & & & & C_{66} \end{bmatrix} \begin{Bmatrix} \varepsilon_{11} \\ \varepsilon_{22} \\ \varepsilon_{33} \\ 2\varepsilon_{23} = \gamma_{23} \\ 2\varepsilon_{31} = \gamma_{31} \\ 2\varepsilon_{12} = \gamma_{12} \end{Bmatrix} \tag{3.3}$$

where $\gamma_{ij} = 2\varepsilon_{ij} (i \neq j)$ is the engineering shear strain.

The material property matrix in Eq. (3.3) generally has 21 independent constants C_{ij} for an anisotropic material with no plane of symmetry, which for an isotropic material reduces to 2:

$$\mathbf{D} = \begin{bmatrix} C_{11} & C_{12} & C_{12} & 0 & 0 & 0 \\ & C_{11} & C_{12} & 0 & 0 & 0 \\ & & C_{11} & 0 & 0 & 0 \\ & & & C_{66} & 0 & 0 \\ & Sym & & & C_{66} & 0 \\ & & & & & C_{66} \end{bmatrix} \tag{3.4}$$

where:

$$C_{66} = \frac{C_{11} - C_{12}}{2} \tag{3.5}$$

The material property matrix D of isotropic material can also be defined based on engineering elastic modulus E and Poisson's ratio v:

$$\mathbf{D} = \frac{E}{(1+v)(1-2v)} \begin{bmatrix} 1-v & v & v & 0 & 0 & 0 \\ & 1-v & v & 0 & 0 & 0 \\ & & 1-v & 0 & 0 & 0 \\ & Sym & & \frac{1-2v}{2} & 0 & 0 \\ & & & & \frac{1-2v}{2} & 0 \\ & & & & & \frac{1-2v}{2} \end{bmatrix} \tag{3.6}$$

Or based on Lame's first and second constants λ and μ:

$$\mathbf{D} = \begin{bmatrix} \lambda + 2\mu & \lambda & \lambda & 0 & 0 & 0 \\ & \lambda + 2\mu & \lambda & 0 & 0 & 0 \\ & & \lambda + 2\mu & 0 & 0 & 0 \\ & & & \mu & 0 & 0 \\ & Sym & & & \mu & 0 \\ & & & & & \mu \end{bmatrix} \tag{3.7}$$

Hence Eq. (3.1) can be reformulated as:

$$\sigma_{ij} = \lambda \delta_{ij} \varepsilon_{kk} + 2\mu \varepsilon_{ij} \tag{3.8}$$

3.4.2 Navier's Equation of Equilibrium and FEM Solution

The momentum equilibrium equation of a finite solid element with mass density ρ subjected to the dynamic load **f** is (Kundu 2004):

$$\sigma_{ij,j} + f_i = \rho \ddot{u}_i \tag{3.9}$$

where time derivatives are denoted by dots over the variable. By substituting Eq. (3.8) in Eq. (3.9), the Navier's equation of elastodynamics in linear elastic, isotropic, loss free and homogeneous medium is obtained as:

$$(\lambda + \mu) u_{j,ji} + \mu u_{i,jj} + f_i = \rho \ddot{u}_i \tag{3.10}$$

or in the vector form:

$$(\lambda + \mu)\nabla(\nabla \cdot \mathbf{u}) + \mu \nabla^2 \mathbf{u} + \mathbf{f} = \rho \ddot{\mathbf{u}} \tag{3.11}$$

where ∇ is the gradient operator vector, and ∇^2 denotes the scalar gradient operator $\nabla \cdot \nabla$. Now for a nonhomogeneous medium with spatial gradient of material properties, the modified Navier's equation is rewritten as:

$$\nabla(\lambda + \mu)\nabla \cdot \mathbf{u} + \nabla . \mu \nabla \mathbf{u} + \mathbf{f} = \rho \ddot{\mathbf{u}} \tag{3.12}$$

Obviously deriving analytical solution for such partial differential equation over a heterogeneous domain is a complicated and practically impossible procedure. Thus FEM solution of equation can be computationally achieved by discretisation of the domain into finite elements and applying equilibrium of momentum over the system (de Borst et al. 2012). The constitutive behaviour of any element is basically determined by its representative nodes α_i, and mapped over local coordinate system by corresponding linear, quadratic or higher order shape functions N_i. In brief, the FEM notation of Eq. (3.12) takes the form:

$$\mathbf{M}\ddot{\mathbf{q}} + \mathbf{K}\mathbf{q} = \mathbf{F} \tag{3.13}$$

$$\mathbf{u} = \mathbf{N}\mathbf{q} \tag{3.14}$$

where **N** is matrix of shape functions, **q** is vector of nodal displacements, and **F** is the vector of nodal external loads of the FEM model. **K** and **M** are FEM stiffness and mass matrices respectively, over the volume of element V as follows:

$$\mathbf{K} = \int_V \mathbf{B}^T \mathbf{D}\mathbf{B} \mathrm{d}V \tag{3.15}$$

$$\mathbf{M} = \int_V \mathbf{N}^{\mathrm{T}} \mathbf{N} \mathrm{d}V \tag{3.16}$$

where **D** is the material property matrix [Eq. (3.4)], and **B** is the strain-nodal displacement matrix:

$$\boldsymbol{\varepsilon} = \mathbf{B}\mathbf{q} \tag{3.17}$$

3.4.3 *The Bloch-Floquet Wave Theory and Unit-Cell's Modal Band Analysis*

According to the Bloch-Floquet wave theory (Kittel 1986) the wave function in a perfectly periodic medium with lattice periodicity vector $\mathbf{A} = \{\, a_x \quad a_y \quad a_z \,\}$ can be defined by **A** periodic modulation of a harmonic function. Therefore, the harmonic oscillations $\mathbf{u} = \{u \quad v \quad w\}^{\mathrm{T}}$ of any spectrum in an infinite PhP lattice over time t and coordination system $\mathbf{x} = \{x \quad y \quad z\}$ can be written as:

$$\mathbf{u}(\mathbf{x}, t) = \tilde{\mathbf{u}}(\mathbf{x}) e^{(i\mathbf{k}\cdot\mathbf{x} - i\omega t)} \tag{3.18}$$

where $\tilde{\mathbf{u}}(x)$ is the **A** periodic function:

$$\tilde{\mathbf{u}}(\mathbf{x}) = \tilde{\mathbf{u}}(\mathbf{x} + \mathbf{A}) \tag{3.19}$$

and $\omega = 2\pi f$ is the circular temporal frequency, $\mathbf{k} = \{k_x \quad k_y \quad k_z\}$ is the circular wave vector (i.e. spatial frequency) and $i = \sqrt{-1}$. Consequently, the following relation between displacements of opposing boundaries can be achieved for any lattice cell:

$$\mathbf{u}(\mathbf{x}, t) = \mathbf{u}(\mathbf{x} + \mathbf{A}, t)e^{i\mathbf{k}\cdot\mathbf{A}} \tag{3.20}$$

By substituting the Bloch-Floquet wave function Eq. (3.18) in Eq. (3.13), the FEM equilibrium equation for any PhP unit-cell under Bloch-Floquet periodic boundary condition [Eq. (3.20)] as a subset of whole lattice structure (and naturally in the absence of external forces) is obtained:

$$\left(-\mathbf{M}\omega^2 + \mathbf{K}(\mathbf{k})\right)\mathbf{q} = 0 \tag{3.21}$$

where stiffness matrix $\mathbf{K}$ is a function of wave vector $\mathbf{k}$ through periodic boundary condition [Eq. (3.20)]. Equation (3.21) can also be written as follows:

$$\left(\mathbf{M}^{-1}\mathbf{K} - \omega^2\mathbf{I}\right)\mathbf{q} = 0 \tag{3.22}$$

where $\mathbf{I}$ is identity matrix. Then modal frequencies ω_i (square root of eigenvalues) and relevant mode shapes (eigenvectors) $\mathbf{q}_i$ can be obtained through eigenvalue analysis of matrix $\mathbf{M}^{-1}\mathbf{K}$ for nontrivial solutions which requires:

$$\left|\mathbf{M}^{-1}\mathbf{K} - \omega^2\mathbf{I}\right| = 0 \tag{3.23}$$

Theoretically the wave vector $\mathbf{k}$ can take any value. However, due to the periodicity of Bloch-Floquet boundary condition [Eq. (3.20)] it can be limited to the Brillouin zone. The Brillouin zone for a square 2D PhP unit-cell with lattice vector $\mathbf{A} = \{a, a, 0\}$ is shown in Fig. 3.4a defining the 2D wave vector space $\mathbf{k} = \{k_x \quad k_y \quad 0\}$ where $\mathbf{k} \in \left[-\frac{\pi}{a}, \frac{\pi}{a}\right]^2$.

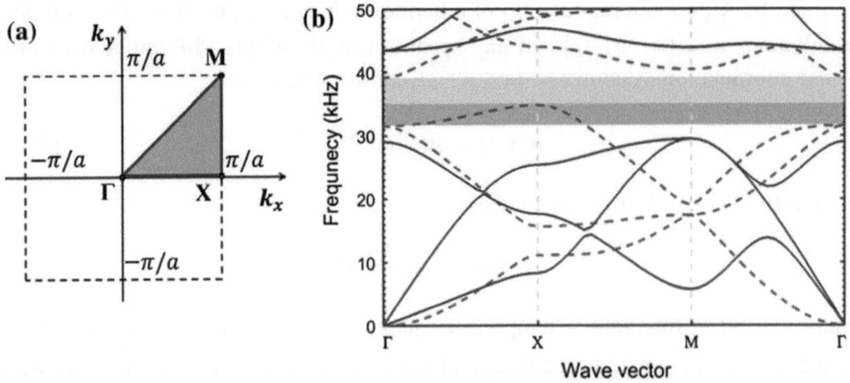

Fig. 3.4 **a** Brillouin zone and irreducible triangular area for 2D PhP with square symmetry, and **b** calculated modal band structure over triangle ΓXM for porous 2D PhP shown in Fig. 1.2

Hence due to the square symmetry of PhP unit-cell, the irreducible triangle ΓXM can be considered only for modal band analysis. However, according to the common practice searching the border of this triangular area merely suffices for definition of modal band structure (Sigmund and Jensen 2003). Likewise, for a 1D PhP with wave vector $\mathbf{k} = \{k\,0\,0\}$ the relevant wave number varies in the irreducible range $k \in \left[0, \frac{\pi}{a}\right]$ over line ΓX in Fig. 3.4.

The calculated modal band structure of a square 2D PhP (Fig. 3.2) at discrete points on irreducible triangle ΓXM is shown in Fig. 3.4b. According to the modal band structure depicted in Fig. 3.4b there is no modal response in the frequency range 34.8–39.2 kHz. However partial bandgaps may exist with bandgap efficiency in particular wave vectors only, like the lower bandgap shaded in the band structure over frequency range 31.6–34.8 kHz.

After calculation of modal band structure over the irreducible Brillouin zone border for specified number of modal frequencies, the bandgap width between two consecutive modal branches j and $j+1$ is defined as:

$$\Delta\omega_j = \omega_{j+1,\min} - \omega_{j,\max} \tag{3.24}$$

where $\omega_{j+1,\min}$ is the minimum angular frequency of upper modal branch $j+1$ and $\omega_{j+1,\max}$ the maximum angular frequency of lower modal branch j, over the n_k discrete wave vectors k_i searched over the Brillouin zone border:

$$\omega_{j+1,\min} = \min_{i=1}^{n_k} \omega_{j+1}(k_i) \tag{3.25}$$

$$\omega_{j,\max} = \max_{i=1}^{n_k} \omega_j(k_i) \tag{3.26}$$

Therefore the relevant frequency RBW is:

$$\Omega_F(j) = \frac{\Delta\omega_j}{\overline{\omega_j}} = \frac{\omega_{j+1,\min} - \omega_{j,\max}}{0.5\left(\omega_{j+1,\min} + \omega_{j,\max}\right)} \tag{3.27}$$

The relative bandgap efficiency may be defined based on the modal frequency ω as given in Eq. (3.27), or based on the eigenvalue of equation of motion which is actually the squared frequency ω^2. So the eigenvalue RBW is:

$$\Omega_E(j) = \frac{\Delta\omega_j^2}{\overline{\omega_j^2}} = \frac{\omega_{j+1,\min}^2 - \omega_{j,\max}^2}{0.5\left(\omega_{j+1,\min}^2 + \omega_{j,\max}^2\right)} \tag{3.28}$$

It is noteworthy that by considering a complex wave vector $\mathbf{k} = \mathbf{k_{Re}} + i\mathbf{k_{Im}}$ the Bloch-Floquet wave function [Eq. (3.18)] becomes:

$$\mathbf{u}(\mathbf{x}, t) = \tilde{\mathbf{u}}(\mathbf{x})e^{(i\mathbf{k_{Re}}\cdot\mathbf{x} - i\omega t)}e^{-\mathbf{k_{Im}}\cdot\mathbf{x}} \tag{3.29}$$

where the last term exponentially decays the wave amplitude over the spatial domain $\mathbf{x}$ by the power of imaginary wave vector $\mathbf{K}_{\mathbf{Im}}$. Thus, considering an imaginary component for the wave vector enables calculation of non-propagating evanescent modes existing within bandgaps with exponential decay through the phononic lattice. However a real wave vector is considered in current study which suffices for determining the bandgaps.

For generality dimensionless modal frequency ω_d is usually studied which is normalised with respect to $C_p = \sqrt{E/\rho}$ and the unit-cell size a:

$$\omega_d = \frac{\omega a}{C_p} \tag{3.30}$$

or

$$f_d = \frac{\omega a}{2\pi C_p} \tag{3.31}$$

C_p is actually the longitudinal wave speed in a thin waveguide. Alternatively, the shear wave speed $C_s = \sqrt{G/\rho}$ may also be used for normalising modal frequency. Unit-cell size of $a = 1$ is conveniently considered for analysis, and then the actual modal band structure for specific lattice periodicity can be defined by scaling the dimensionless frequency.

For modal band analysis of 1D PhP a specialised FEM code is developed in MATLAB. However, ANSYS APDL FEM solver is employed for modal analysis of 2D PhPs due to its computational speed and finite element modelling features particularly in evaluating introduced secondary objectives. For this purpose, macro scripts are coded for automated parametric analysis of randomly generated topologies during GA optimisation.

3.5 Concluding Remarks

Gradient based and stochastic optimisation algorithms were briefly introduced in this chapter and their application for structural topology optimisation was discussed. The main advantages and disadvantages of both approaches were addressed and GA was concluded to be a suitable stochastic optimisation algorithm for the scope of present study. Then non-dominated sorting GA (NSGA-II) was introduced as the fundamental optimisation framework to be implemented for multi-objective topology optimisation of PhPs, and the basic steps of optimisation algorithm were defined. Afterwards, the definition of binary design variable domain for 1D bi-material and 2D porous PhPs and relevant topology mapping were explained.

The constitutive stress-strain formulation of the elastic base material and Navier's equation of equilibrium of heterogeneous PhP medium were also given. Modal band analysis of PhP unit-cell, as the common objective of all research stages, and relevant boundary conditions and FEM formulation were described. Detailed formulation and evaluation procedure of other objectives will be discussed later on, in relevant chapters.

References

Bilal, O. R., & Hussein, M. I. (2011). Ultrawide phononic band gap for combined in-plane and out-of-plane waves. *Physical Review E, 84*(6), 065701.

Cavazzuti, M. (2013). *Optimization methods: From theory to design scientific and technological aspects in mechanics*. Berlin: Springer.

de Borst, R., Crisfield, M. A., Remmers, J. J. C., & Verhoosel, C. V. (2012). *Non-linear finite element analysis of solids and structures* (2nd ed.). Chichester: Wiley.

Fan Li, Y., Huang, X., Meng, F., & Zhou, S. (2016). Evolutionary topological design for phononic band gap crystals. *Structural and Multidisciplinary Optimization*, 1–23.

Gao, H., Matsumoto, T., Takahashi, T., & Isakari, H. (2013). Analysis of band structure for 2D acoustic phononic structure by BEM and the block SS method. *CMES—Computer Modeling Engineering & Sciences, 90,* 283–301.

Gazonas, G. A., Weile, D. S., Wildman, R., & Mohan, A. (2006). Genetic algorithm optimization of phononic bandgap structures. *International Journal of Solids and Structures, 43*(18–19), 5851–5866.

Guest, J. K., Prévost, J., & Belytschko, T. (2004). Achieving minimum length scale in topology optimization using nodal design variables and projection functions. *International Journal for Numerical Methods in Engineering, 61*(2), 238–254.

Guest, J. K., & Smith Genut, L. C. (2010). Reducing dimensionality in topology optimization using adaptive design variable fields. *International Journal for Numerical Methods in Engineering, 81*(8), 1019–1045.

Huang, X., & Xie, M. (2010). *Evolutionary topology optimization of continuum structures: Methods and applications*. Chichester: Wiley.

Hussein, M. I., Hamza, K., Hulbert, G. M., & Saitou, K. (2007, November). Optimal synthesis of 2D phononic crystals for broadband frequency isolation. *Waves in Random and Complex Media, 17*(4), 491–510.

Hussein, M. I., Hamza, K., Hulbert, G. M., Scott, R. A., & Saitou, K. (2006, January 1). Multiobjective evolutionary optimization of periodic layered materials for desired wave dispersion characteristics. *Structural and Multidisciplinary Optimization,* 31(1), 60–75.

Kafesaki, M., & Economou, E. N. (1999). Multiple-scattering theory for three-dimensional periodic acoustic composites. *Physical Review B, 60*(17), 11993.

Kao, C., Osher, S., & Yablonovitch, E. (2005). Maximizing band gaps in two-dimensional photonic crystals by using level set methods. *Applied Physics B, 81*(2–3), 235–244.

Kittel, C. (1986). *Introduction to solid state physics*. New York: Wiley.

Kundu, T. (2004). *Ultrasonic nondestructive evaluation—Engineering and biological material characterization*. Boca Raton: CRC Press.

Liu, Z.-f., Wu, B., & He, C.-f. (2014). Band-gap optimization of two-dimensional phononic crystals based on genetic algorithm and FPWE. *Waves in Random and Complex Media*, no. ahead-of-print, pp. 1–20.

Meng, F., Huang, X., & Jia, B. (2015). Bi-directional evolutionary optimization for photonic band gap structures. *Journal of Computational Physics, 302,* 393–404.

Pratap, A., Agarwal, S., & Meyarivan, T. (2002). A fast and elitist multiobjective genetic algorithm: NSGA-II. *IEEE Transactions on Evolutionary Computation, 6*(2), 182–197.

Radman, A. (2013). *Bi-directional evolutionary structural optimization (BESO) for topology optimization of material's microstructure*. RMIT University.

Rennera, G., & Ekart, A. (2003). Genetic algorithms in computer-aided design. *Computer-Aided Design, 35,* 709–726.

Sigmund, O., & Jensen, J. S. (2003). Systematic design of phononic band-gap materials and structures by topology optimization. *Philosophical Transactions of the Royal Society of London A, 361,* 1001–1019.

Sigmund, O. (2011). On the usefulness of non-gradient approaches in topology optimization. *Structural and Multidisciplinary Optimization, 43*(5), 589–596.

Svanberg, K. (1987). The method of moving asymptotes—A new method for structural optimization. *International Journal for Numerical Methods in Engineering, 24*(2), 359–373.

Tsai, S. W. (1992). *Theory of composites design,* Think composites Dayton.

Wang, M. Y., Wang, X., & Guo, D. (2003, January 3). A level set method for structural topology optimization. *Computer Methods in Applied Mechanics and Engineering, 192*(1–2), 227–246.

Wang, Y., Gao, J., Luo, Z., Brown, T., & Zhang, N. (2016a). Level-set topology optimization for multimaterial and multifunctional mechanical metamaterials. *Engineering Optimization,* 1–21.

Wang, Y., Luo, Z., Zhang, N., & Wu, T. (2016b). Topological design for mechanical metamaterials using a multiphase level set method. *Structural and Multidisciplinary Optimization,* 1–16.

Yan, Z.-Z., & Wang, Y.-S. (2006). Wavelet-based method for calculating elastic band gaps of two-dimensional phononic crystals. *Physical Review B, 74*(22), 224303.

Chapter 4
Optimisation of Bi-material Layered 1D Phononic Crystal Plates (PhPs)

4.1 Introduction

The commonly studied type of 1D periodic layered PhPs is the one made by periodic insertion of a centric scattering layer in a base plate. Chen et al. (2006) investigated the bandgaps of low order Lamb waves in 1D PhPs made by periodic insertion of tungsten strips in silicon background. The effect of unit-cell's aspect ratio to the bandgap width was studied for specified filling fraction of 0.5, and maximum bandgap was obtained for aspect ratio (width to thickness) 0.53. Gao et al. (2008) also studied the bandgap performance of such 1D PhPs when coated on a thin uniform substrate plate with soft or stiff constitutive materials. Similarly the bandgap efficiency of such 1D PhP sandwiched between two identical uniform plates was explored by Chen and Han (2010).

Although considering centric scattering inclusions in the base plate produces a PhP with acoustic bandgaps, the optimised arrangement of scattering layers can be defined by inverse design of PhP unit-cell in order to obtain wide bandgaps of guided wave modes of interest. As discussed in Chap. 2, the topology optimisation of 1D layered PhCrs have been well studied for bandgaps of in-plane and anti-plane bulk waves. However, to the best of the Author's knowledge based on the comprehensive literature review, there has been no study on optimum topology of PhPs for maximisation of bandgaps of guided waves.

Moreover it is worthwhile to study the optimum topology of PhP unit-cells for maximised specific bandgap width as a function of filling fraction of inclusions. High contrast of constitutive materials is vital for opening bandgaps and subsequently adjusting fraction of soft or stiff inclusions is applicable e.g. to satisfy weight or production requirements. Also multiscale modulation of material properties by gradient of PhCr unit-cell throughout the lattice array can produce a multifunctional structure which has additional functionality at scales larger than PhCr unit-cell size. Followings are potential examples of application of gradient PhP concept:

S. Hedayatrasa, *Design Optimisation and Validation of Phononic Crystal Plates for Manipulation of Elastodynamic Guided Waves*, Springer Theses,
https://doi.org/10.1007/978-3-319-72959-6_4

- Designing a vibroacoustic device by which the frequencies within bandgap are highly attenuated and at the same time those below bandgap are steered parallel to the plate's free surfaces. The steering efficiency depend on the variation rate of effective refraction index (i.e. filling fraction) along the PhP.
- A vibroacoustic transducer may be designed which filters out high frequency contents and resonates at intended low frequency.
- In structural concept, the distribution of heavy-stiff constituent may be designed so that operational loads are minimised and/or fundamental modal frequencies are shifted.

Hence, this chapter is dedicated to:

1. Investigating the optimised topology of PhP unit-cell for maximised RBW of in-plane guided (Lamb) waves with respect to filling fraction of constituents.
2. Exploring relative bandgap efficiency of gradient PhPs comprised of consecutive unit-cells of different filling fractions with optimised topology.

The governing equations and implemented FEM-GA basic formulation are firstly described in Sect. 5.2. The results of developed FEM code for dispersion analysis of single material plate along with band structure analysis of bi-material plate unit-cells are delivered and discussed in Sects. 5.3. Then topology optimisation results are presented in Sect. 5.4 and definite prescribed topologies are introduced and evaluated premised on achieved optimised topologies. Finally, the frequency response of representative gradient PhP structures comprised of a variety of unit-cell topologies is investigated in Sect. 5.5.

4.2 Theory and Constitutive Formulation

4.2.1 Wave Propagation in 1D Periodic PhP

The focus of this research is to study the dispersion and modal band structure of elastic waves in longitudinally periodic PhP with no material gradient through the thickness. A schematic of such a 1D periodic plate with thickness *h* and its basic unit-cell with periodicity length *a* (along *x* direction) is shown in Fig. 4.1.

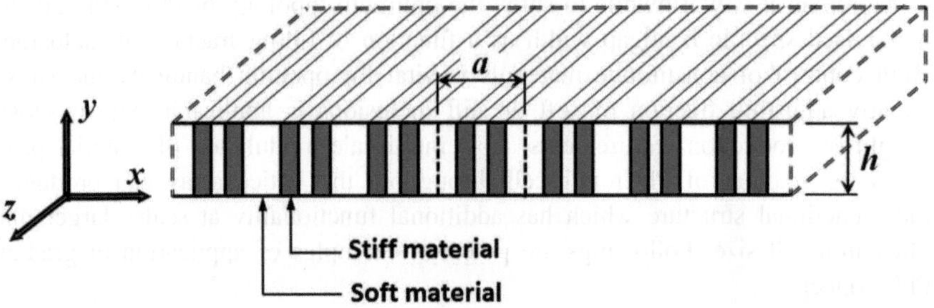

Fig. 4.1 Schematic of a 1D PhP along x direction with lattice constant *a* and thickness *h*

Relatively large width (along z-axis) is considered in which the effect of free edges (at $z+$ and $z-$) can be neglected and so by assumption of uniform displacement field in z-axis ($\partial/\partial z = 0$) the problem is confined to xy-plane.

Based on Navier's equation of equilibrium, two sets of independent equations can be developed for guided wave propagation in such heterogeneous plate: a couple of equations for in-plane (Lamb) waves:

$$\rho \frac{\partial^2 u}{\partial t^2} = \frac{\partial}{\partial x}\left(C_{11}\frac{\partial u}{\partial x} + C_{12}\frac{\partial v}{\partial y}\right) + C_{44}\left(\frac{\partial^2 u}{\partial y^2} + \frac{\partial^2 v}{\partial x \partial y}\right) \tag{4.1}$$

$$\rho \frac{\partial^2 v}{\partial t^2} = C_{11}\frac{\partial^2 v}{\partial y^2} + C_{12}\frac{\partial^2 u}{\partial x \partial y} + \frac{\partial}{\partial x}\left(C_{44}\left(\frac{\partial v}{\partial x} + \frac{\partial u}{\partial y}\right)\right) \tag{4.2}$$

and an equation for anti-plane SH modes:

$$\rho \frac{\partial^2 w}{\partial t^2} = \frac{\partial}{\partial x}\left(C_{44}\frac{\partial w}{\partial x}\right) + C_{44}\left(\frac{\partial^2 w}{\partial y^2}\right) \tag{4.3}$$

The general Bloch-Floquet solution for the assumed 1D PhP with lattice constant a is defined by harmonic modulation of an a periodic field, as follows:

$$\mathbf{u}(x, y, t) = \tilde{\mathbf{u}}(x, y)e^{i(\mathrm{kx} - \omega t)} \tag{4.4}$$

$$\tilde{\mathbf{u}}(x, y) = \tilde{\mathbf{u}}(x + a, y) \tag{4.5}$$

where $k = k_x$ is spatial angular frequency (wave number) in x-axis. So the relevant infinitesimal complex strain matrices for in-plane and anti-plane waves respectively are:

$$\begin{bmatrix} \varepsilon_x & \varepsilon_y & \gamma_{xy} \end{bmatrix}^T = \begin{bmatrix} \frac{\partial \tilde{u}}{\partial x} + ik\tilde{u} & \frac{\partial \tilde{v}}{\partial y} & \frac{\partial \tilde{u}}{\partial y} + \frac{\partial \tilde{v}}{\partial x} + ik\tilde{v} \end{bmatrix}^T e^{i(kx - \omega t)} \tag{4.6}$$

$$\begin{bmatrix} \gamma_{zx} & \gamma_{zy} \end{bmatrix}^T = \begin{bmatrix} \frac{\partial \tilde{w}}{\partial x} + ik\tilde{w} & \frac{\partial \tilde{w}}{\partial y} \end{bmatrix}^T e^{i(kx - \omega t)} \tag{4.7}$$

where $\tilde{\mathbf{u}} = \{\tilde{u}\ \tilde{v}\ \tilde{w}\}$. Periodic plates made by silicon (Si) and tungsten (W) strips are modelled and studied in this research with material properties listed in Table 4.1.

Table 4.1 Material properties used for FEM modelling of 1D PhP (Chen et al. 2006)

Material	Silicon (Si)	Tungsten (W)
ρ (kg/m^3)	2330	19,200
C_{11} (GPa)	165.7	502
C_{12} (GPa)	63.9	199
C_{44} (GPa)	79.6	152
E (GPa)	130	389
v	0.278	0.284
$C_p = \sqrt{E/\rho}$ (m/s)	7470	4501

Silicon and tungsten are of potential matrix and scattering materials respectively, for fabrication of phononic bandgaps and their micro-fabricated samples have been produced by Sandia National Lab, USA (Olsson III and El-Kady 2009; Olsson III et al. 2008).

4.2.2 Specialised FEM Model

A specialised FEM model using first order quadrilateral elements is developed in MATLAB (*MATLAB 8.0, The MathWorks Inc.*) for dispersion analysis of guided waves and calculation of the band structure of hypothesized periodic plate (*not by using MATLAB FEM toolbox*). The Bloch-Floquet wave function is embedded in stiffness matrix through Eqs. (4.6) and (4.7) instead of being solely applied through harmonic boundary conditions. This enables calculation of unfolded modal band structure of guided waves with respect to actual (non-periodic) wave number outside the Brillouin zone.

Taking into account Bloch-Floquet solution (Eq. 4.4), the FEM notation of displacement vector and equation of motion for a loss free elastic model with stiffness matrix **K** and mass matrix **M** become:

$$\mathbf{u} = \mathbf{N}\tilde{\mathbf{q}}e^{i(kx-\omega t)} \tag{4.8}$$

$$(\mathbf{K}(k) - \mathbf{M}\omega^2)\tilde{\mathbf{q}}e^{i(kx-\omega t)} = 0 \tag{4.9}$$

where $\tilde{\mathbf{q}}$ is the vector of nodal displacements corresponding to periodic term of displacement field $\tilde{\mathbf{u}}$. With regard to Eqs. (4.6) and (4.7) the strain-displacement matrix and so stiffness matrix **K** are complex and dependent on longitudinal wave number k. Due to periodicity of structure just a representative unit-cell with length equal to lattice constant a is modelled. Periodic boundary condition is therefore applied by equalising corresponding periodic degrees of freedom at the left and right edges of the model. The top and bottom traction free edges are left with natural free boundary condition. Finally, the modal response of unit-cell is obtained by eigenvalue analysis of Eq. (4.10) for non-trivial solutions:

$$\left|\mathbf{K}_r(k) - \mathbf{M}_r\omega^2\right| = 0 \tag{4.10}$$

where $\mathbf{K}_r$ and $\mathbf{M}_r$ are reduced mass and stiffness matrices obtained after applying periodic boundary condition through master-sleeve transform (Felippa 2004). So the gradient of eigenvalues versus wave number k is determined over irreducible Brillouin zone $0 \leq k \leq \pi/a$, giving the unidirectional band structure of 1D PhP unit-cell.

Appropriate mesh resolution should be defined in FEM simulation so that the tiny wave oscillations of highest frequency of interest are approximated adequately. According to the common practice, 10–20 nodes have to be considered over the wavelength (Gopalakrishnan et al. 2007). Hence mesh density with at least 15 nodes per shortest wavelength is ensured based on lower elastodynamic wave speed which pertains to the tungsten inclusion in this study (Table 4.1).

4.2.3 Objective Functions and Optimisation Algorithm

This study is concerned with finding optimum topology of 1D PhP unit-cell with respect to filling fraction of stiff inclusion for maximising the bandgap width of Lamb waves.

As for the first objective, as discussed in Chap. 3, the fundamental requirement is to open widest bandgaps at lowest potential frequency range for specific plate's thickness at prescribed lattice constant. This way the widest phononic controllability frequency range is achieved through smallest unit-cell size compared with the wavelength. So the first objective function F_1 to be maximised is formulated as the sum of eigenvalue RBWs (Eq. 3.28) within the desired modal branches j_1 to j_2:

$$F_1 = \sum_{j=j_1}^{j_2} \Omega_E(j) \tag{4.11}$$

The second objective F_2 to be minimised is then defined as absolute deviation of filling fraction of stiff tungsten inclusion v_f from its prescribed reference value v_r:

$$F_2 = \left| v_f - v_r \right| \tag{4.12}$$

The filling fraction itself may be introduced as the second objective to be minimised or maximised. However, the objective function defined in Eq. (4.12) is preferably considered to explore optimised set of topologies around various filling fractions. In any optimisation, optimised topologies concerning other filling fractions with higher bandgap efficiency also appear in Pareto front. Due to stochastic nature of implemented optimisation algorithm, optimised topologies of specific filling fraction obtained through different optimisations are not necessarily the same.

The basic steps of multi-objective optimisation algorithm implemented herein are as follows:

1. Create a random bit-string initial population for the bi-material problem defining the material type at each element of FEM model as a design variable (gene). Verify the fitness of individuals, define their rank and crowding distance and sort them.
2. Select individuals to be considered as parents of next generation through tournament selection. Two individuals are selected at random and the one with highest rank is selected as parent. If both have the same rank, the one with highest crowding distance is selected.

3. Create offspring population by applying single point crossover with certain probability on couples of randomly selected parents providing two Children per couple.
4. Mutate gens of offspring population with specified probability.
5. Produce intermediate population by accumulating offspring and current population and sort them based on ranking and crowding distance.
6. Select individuals for next generation, first based on their ranking and second, if the first rank is bigger than required population, based on their crowding distance.
7. Go to step-2, stop when the desired maximum number of generations is reached.

4.3 Guided Waves Dispersion and Bandgaps

The modal band structure of guided waves in 1D silicon-tungsten PhP unit-cells with regular centric topology is studied in this section for various filling fractions and different aspect ratios. Initially, the accuracy and competency of developed FEM model in determining the guided wave dispersion and frequency response of structures is evaluated. Multilayer FEM model of a pure silicon square cell with unit length as shown in Fig. 4.2 is employed for this purpose. 50 element divisions through thickness are considered to determine the unfolded dispersion curves of guided waves including first 10 Lamb wave modes. Whereas linear quadratic elements are implemented, applying Bloch-Floquet periodic boundary condition on two edges of single longitudinal element makes the periodic component of displacement vector $\tilde{\mathbf{u}}$ independent of x over the entire unit-cell length. Consequently, pure harmonic oscillations of guided waves along a single material plate are simulated through this multilayer model.

Figure 4.3a shows the wave dispersion in terms of phase velocity versus the product of frequency and plate's thickness, comparable to the reference analytical results by (Degertekin et al. 1995). Excellent agreement was observed by comparison of these two. Moreover, the dimensionless modal band structure, as

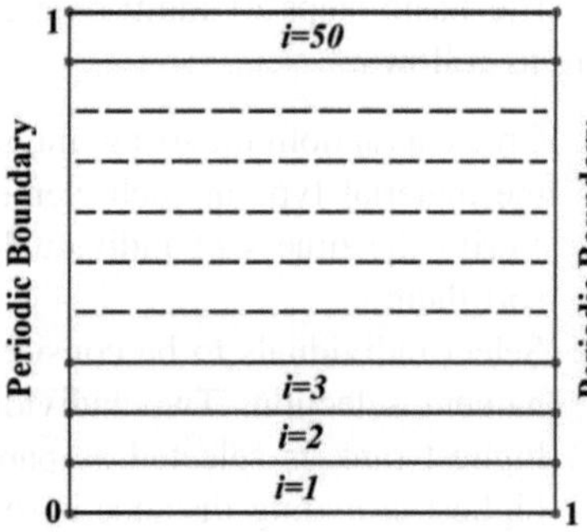

Fig. 4.2 Multilayer FEM model of a plate unit-cell for guided waves dispersion analysis, having periodic boundary at left and right edges and 50 element divisions evenly through thickness

displayed in Fig. 4.3b, is determined based on dimensionless wave number $k_d = ka/\pi$ and dimensionless angular frequency $\omega_d = \omega h/C_p$, where $C_p = 7470$ m/s (Table 4.1).

In the modal band structures, solid lines stand for in-plane (Lamb) modes and dash lines for anti-plane SH modes. The colours differentiate various frequency modes, however Lamb modes intersect somewhere between adjacent in-plane modes as highlighted by bold dots in Fig. 4.3 for symmetric modes S_1 and S_2. This is investigated by identifying symmetry of modal shapes at different wavelengths.

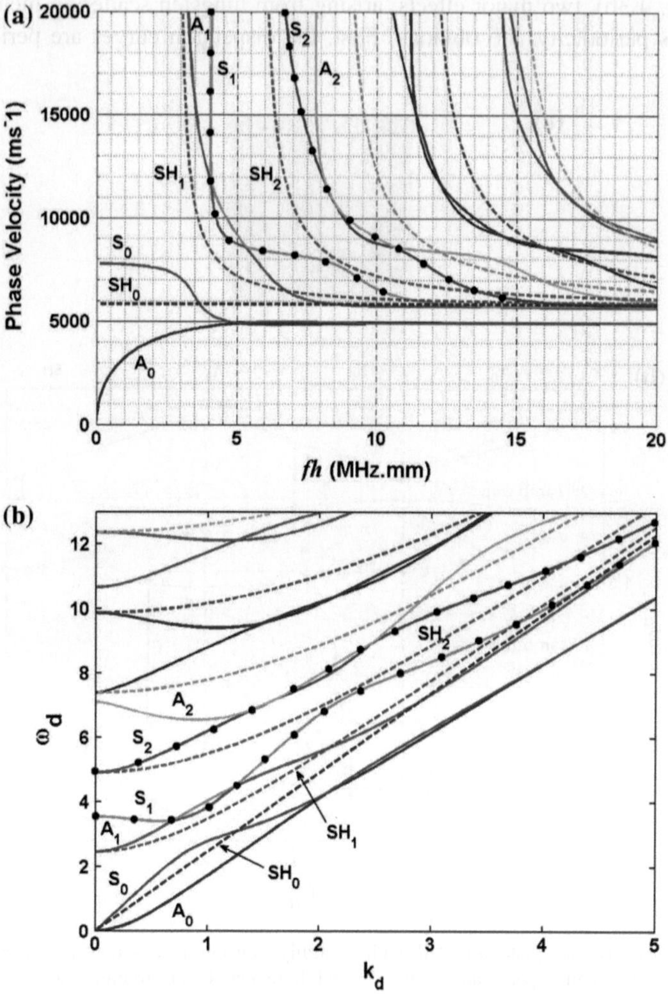

Fig. 4.3 Guided wave dispersion curves obtained for a pure silicon plate; **a** Phase velocity versus product of frequency and plate's thickness *fh*, **b** non-dimensional modal band structure

In order to study the band structure of guided waves in 1D periodic plate structures, primarily a regular bi-material PhP unit-cell with centrally concentrated scattering inclusion shown in Fig. 4.4a is modelled. The aspect ratio of unit-cell $a/h = 2$ and filling fraction of tungsten scattering phase $v_f = 0.5$. The non-dimensional modal band structure of this unit-cell is then calculated as shown in Fig. 4.4b.

Solid lines define the in-plane (Lamb) modes and dash lines the anti-plane SH modes. Dispersion of different modal frequencies is highlighted by colours, and intersections of symmetric (S) and asymmetric (A) Lamb modes are defined by appropriate markings on first 6 in-plane modes. By comparing the dispersion curves of this silicon-tungsten periodic plate unit-cell (Fig. 4.4b) with that of pure silicon plate (Fig. 4.3b), two major effects, arising from tungsten scattering inclusion and structure's periodicity, are obvious. First, the dispersion curves are periodic with

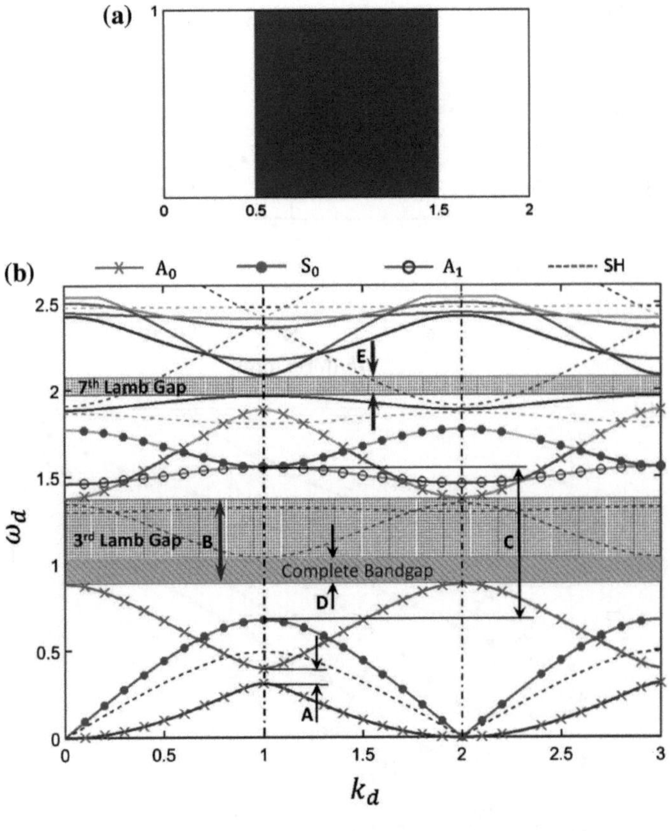

Fig. 4.4 **a** Bi-material plate unit-cell with centrally concentrated scattering inclusion, filling fraction $v_f = 0.5$ and aspect ratio $a/h = 2$, and **b** related band structure of first few modes including a complete bandgap and two Lamb gaps; dash lines stand for anti-plane SH modes, and solid lines stand for Lamb modes on which low order symmetric and asymmetric Lamb modes are highlighted by different markings

respect to the wave number and symmetric about $k_d = 1$ in the range k_d = 0–2. This is due to periodicity of Bloch-Floquet boundary condition and appearance of modal branches folded back to the first Brillouin zone. Second, frequency bandgaps open in the band structure of periodic plate unit-cell in which some or all wave modes are not feasible. Bandgap A is the first bandgap exclusive to principal asymmetric Lamb mode (A_0) as principal symmetric mode (S_0) exists in this range. Bandgap B is a bandgap of mixed Lamb modes (modes 3rd and 4th) and also the second exclusive bandgap of A_0. Although no Lamb mode appears in the frequency range of bandgap B, two SH modes interrupt it. Hence, bandgap D is the only complete bandgap in which no guided wave is feasible. Bandgap C is also the first exclusive bandgap of principal symmetric mode S_0. The second Lamb gap observed among displayed modes is bandgap E between Lamb modes 7th and 8th which is interrupted by one SH mode. The focus of this research is to study complete bandgaps of in-plane modes which stop propagation of low order Lamb waves regardless of anti-plane SH modes, e.g. bandgap B and bandgap E in Fig. 4.4b.

The results displayed in Fig. 4.4 show the band structure of a PhP unit-cell with specific aspect ratio and filling fraction. So the existence and variation of bandgaps at different filling fractions and aspect ratios are further investigated. First 10 Lamb gaps which are likely to open between subsequent pairs of first eleven in-plane modes are studied, orderly named 1–10 (e.g. bandgap B in Fig. 4.4b that is 3rd Lamb gap opened between 3rd and 4th in-plane modes, and bandgap E that is 7th Lamb gap opened between 7th and 8th in-plane modes). This enables investigation of broad bandgap efficiency of PhP unit-cell.

Figure 4.5 presents the variation of frequency RBW Ω_F of the two major gaps 3rd and 7th (Fig. 4.4b) versus aspect ratio of plate unit-cell. The filling fraction is 0.5 and the length of unit-cell with fixed unit thickness varies to change its aspect

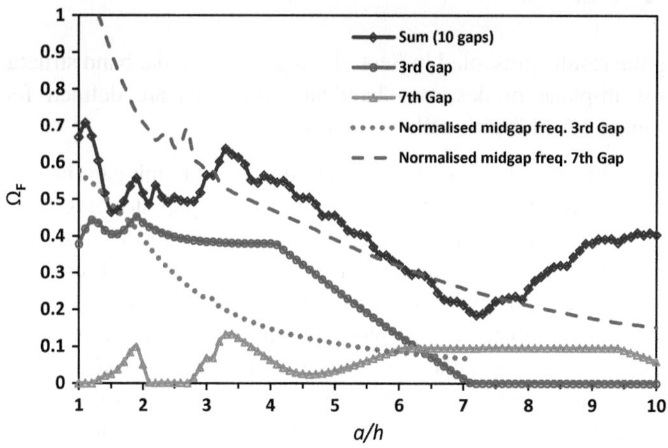

Fig. 4.5 Variation of RBW of 3rd, 7th Lamb gaps and sum of first 10 likely gaps as well as gradient of normalised midgap frequencies versus aspect ratio of regular PhP unit-cell with centric scattering inclusion, constant thickness and $v_f = 0.5$

ratio. Total relative bandgap of all 10 likely gaps opened within this range is also delivered showing the contribution of other gaps and the broad bandgap efficiency of the PhP unit-cell. Accordingly, maximum total relative gap occurs at $a/h = 1.1$ while for the 3rd and 7th gap it happens at $a/h = 1.9$. The two dash lines plot the gradient of midgap frequency of 3rd and 7th gap versus aspect ratio, confirming lower frequency gap at higher aspect ratios. This is justified by bigger feature size of unit-cells at higher aspect ratios (for specified thickness) that can affect larger wavelengths carrying lower frequencies. At high ratios the 3rd gap diminishes, however 7th gap remains and rises slightly.

The total RBW climbs after around $a/h = 7$ where the 3rd band vanishes, explaining significant contribution of other gaps in this range. For more thorough understanding, the opening and variation of all 10 likely bandgaps at a set of aspect ratios 1, 2, 4 and 10 versus filling fraction are scrutinised and presented in Fig. 4.6. The left side figures show gradient of bandgap widths for the thickness $h = 1$ mm. The right side ones show gradient of RBW of first major gap as well as total RBW of all gaps. At aspect ratio $a/h = 1$ the 3rd and 6th gaps are noticeable. At aspect ratio $a/h = 2$ the 3rd gap is outstanding over the range of filling fractions and therefore the total RBW follows the 3rd gap and have low difference with it. As the aspect ratio rises, more gaps open and grow leading to considerably higher total bandgap compared to the 3rd gap. Eventually at $a/h = 10$ the 3rd gap is absent and the 4th gap stands as the first major gap. The 3rd gap has widest bandgap size at filling fraction of around $v_f = 0.4$–0.6, however the total bandwidth and other gaps have multiple extremes at different filling fractions. The 1st gap does not open at all and the 2nd gap is minor at all cases.

4.4 Topology Optimisation

Following the results presented in Sect. 4.3 regarding modal band structure analysis of few first in-plane modes, two bandgap objectives are defined for topology optimisation of 1D PhP unit-cells as follows:

- Bandgap Objective-1: Maximising RBW of the 3rd Lamb gap (between in-plane modes 3rd and 4th) as the major low frequency gap stopping propagation of principal symmetric and asymmetric Lamb wave modes (Fig. 4.4b).
- Bandgap Objective-2: Maximising the total RBW of 10 Lamb gaps (if any) within the first 11 in-plane modes prohibiting propagation of low order lamb waves. Bandgap Objective-2 explores the feasibility of maximising Lamb gap width over a wide frequency range and achieve broadband phononic controllability. Existence and significance of Lamb gaps opening in this range, for specific case of centric PhP topology, were studied in previous section. This is an example of topology optimisation for broadband multimodal bandgap, and the number of modal branches included in optimisation maybe reduced or increased for a specific application.

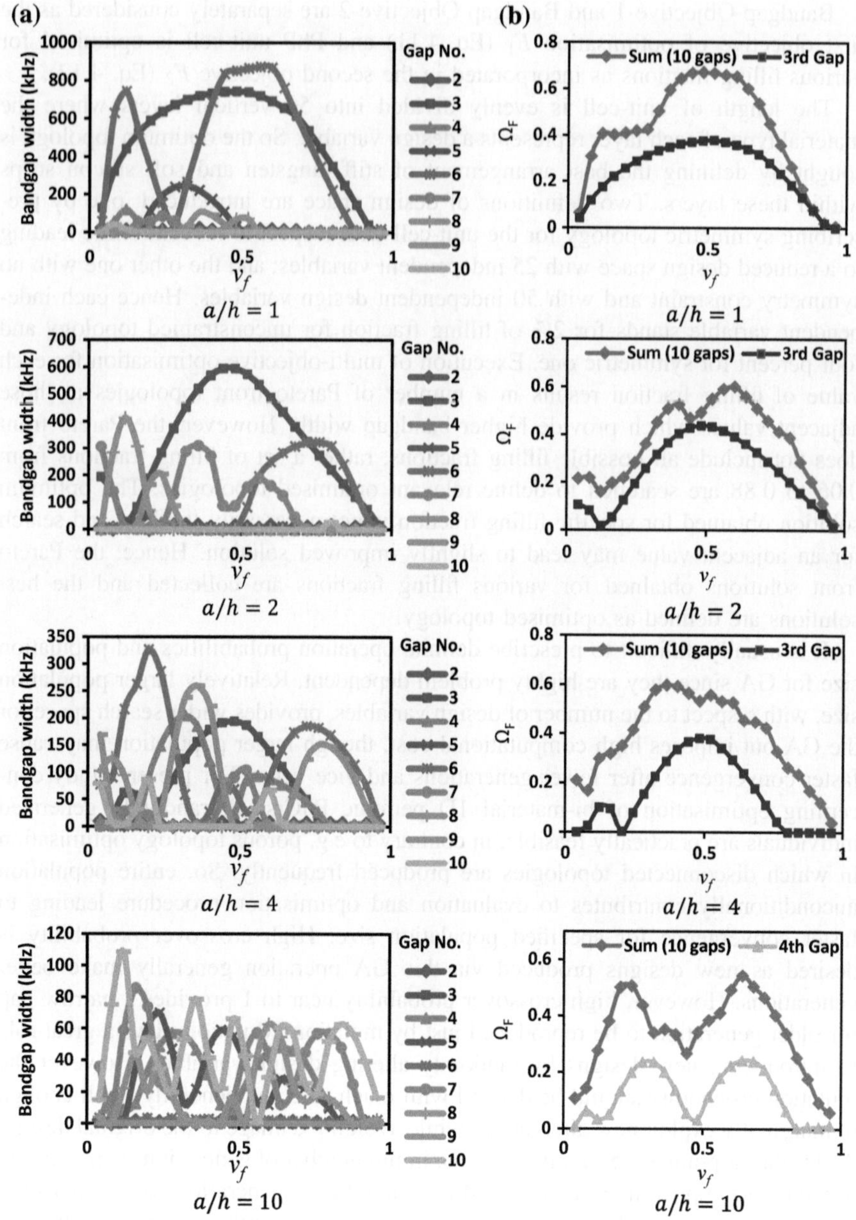

Fig. 4.6 **a** Bandwidth of first 10 likely Lamb gaps for thickness 1 mm and **b** RBW of first major Lamb gap and total RBW of first 10 likely Lamb gaps, versus filling fraction of tungsten inclusion v_f at different aspect ratios

Bandgap Objective-1 and Bandgap Objective-2 are separately considered as the first objective of optimisation F_1 (Eq. 4.11) and PhP unit-cell is optimised for various filling fractions as incorporated in the second objective F_2 (Eq. 4.12).

The length of unit-cell is evenly divided into 50 vertical layers where the material type of each layer represents a design variable. So the optimum topology is sought by defining the best arrangement of stiff tungsten and soft silicon strips within these layers. Two definitions of design space are introduced: one by prescribing symmetric topology for the unit-cell with respect to its centre line leading to a reduced design space with 25 independent variables; and the other one with no symmetry constraint and with 50 independent design variables. Hence each independent variable stands for 2% of filling fraction for unconstrained topology and four percent for symmetric one. Execution of multi-objective optimisation for each value of filling fraction results in a number of Pareto front topologies at those adjacent values which provide higher bandgap width. However, the Pareto front does not include all possible filling fractions, rather a set of filling fractions from 0.06 to 0.88 are searched to define relevant optimised topologies. The optimum solution obtained for specific filling fraction is not necessarily the best and search for an adjacent value may lead to slightly improved solution. Hence, the Pareto front solutions obtained for various filling fractions are collected and the best solutions are defined as optimised topology.

It is usually difficult to prescribe definite operation probabilities and population size for GA since they are highly problem dependent. Relatively larger population size, with respect to the number of design variables, provides wider search space for the GA but imposes high computational cost, though larger population may cause faster convergence after fewer generations and vice versa. For present study concerning optimisation of bi-material 1D periodic PhCrs, all randomly generated individuals are practically feasible, in contrary to e.g. porous topology optimisation in which disconnected topologies are produced frequently. So, entire population unconditionally contributes to evaluation and optimisation procedure leading to faster convergence for specified population size. High crossover probability is desired as new designs produced via this GA operation generally make better generations. However, high crossover probability near to 1 provides a narrow gap for older generation to be reproduced just by mutation. Mutation plays a great role in introducing new designs by randomly altering design variables. However the mutation probability should be defined with much care and generally small enough to marginally exploit new alternatives while avoiding damage to the original design. As for the stopping criteria, often the maximum number of generations is prescribed in which no improvement is observed for a number of generations before reaching the maximum limit. The GA parameters chosen for optimisation execution of present work are listed in Table 4.2.

In order to verify each individual topology in GA population, 15 points along the first irreducible Brillouin zone are searched. However for calculating the frequency band structure and RBW of achieved topologies, 50 points are searched for higher accuracy.

Table 4.2 GA parameters used for multi-objective optimisation of 1D PhP unit-cell

Design space	Design variables	Population size	Generations	Crossover probability	Mutation probability
Symmetric	25	75	40	0.95	0.02
Unconstrained	50	120	80	0.95	0.02

4.4.1 Bandgap Objective-1

Initially unit-cell with aspect ratio $a/h = 2$ is studied showing high RBW in regular centric topology (Fig. 4.5). The Pareto front and relevant convergence history for the case of reference filling fraction $v_f = 0.12$ are shown in Fig. 4.7. Figure 4.7a, b illustrate optimisation of symmetric topology while Fig. 4.7c, d depict optimisation of unconstrained topology.

The number of solutions achieved from optimisation with unconstrained topology is more than that of symmetric one. Actually, the symmetric constraint decreases the diversity of design space and provides solutions for filling fraction steps of 0.04. Also the number of Pareto front solutions varies based on the

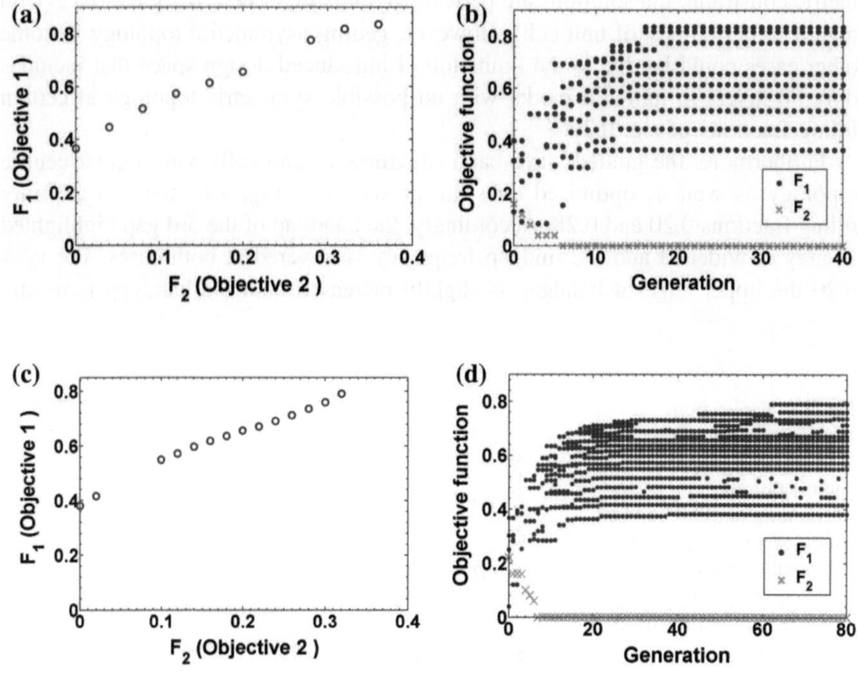

Fig. 4.7 Pareto front and convergence history of GA optimisation for PhP unit-cell with $a/h = 2$ and maximised Bandgap Objective-1 at $v_f = 0.12$, **a**, **b** symmetric topology, **c**, **d** unconstrained topology

individuals sorted in the first rank, and the non-dominated results pertaining to other filling fractions may be located in higher ranks. The algorithm converged to the topology with desired filling fraction $v_f = 0.12$ (i.e. $F_2 = 0$) after 7 and 8 generations for symmetric and unconstrained optimisations respectively. Afterwards, the Pareto solution with lowest bandgap objective F_1 relates to this filling fraction which converges at early stages of optimisation (Fig. 4.7b, d).

The RBW of 3rd gap for all obtained Pareto front solutions for various filling fractions is then calculated and shown in Fig. 4.8. Moreover, the RBW of regular centric topologies versus filling fraction is inserted in Fig. 4.8 to somehow verify the efficiency of obtained optimised topologies.

By comparing the results of topology optimisation with regular centric topology, obviously the RBW is increased in the range $0.06 < v_f \leq 0.4$ and converges to the centric topology for $0.4 < v_f$. Moreover, the bandgap widths of the best solutions from both symmetric and unconstrained topologies are close, although unconstrained solution provides higher RBW at $v_f = 0.08$.

The best optimised topologies which have increased bandgap width compared to regular centric topologies are collected from Pareto front solutions and presented in Fig. 4.9a, b for symmetric and unconstrained cases respectively. The unconstrained topologies are rearranged so that the scattering part is concentrated at the right hand of unit-cell. Regarding these unconstrained topologies, despite having no symmetric constraint, the solutions are perfectly symmetric at $v_f = 0.28$ and 0.32 (with regard to periodicity of unit-cell). However, getting asymmetric topology in some other cases could be due to the limitation of introduced design space that includes only 50 layers in unit-cell model with no possible symmetric topology at certain filling fractions of e.g. 0.26.

Furthermore, the guided wave band structures of unit-cells with regular centric topology as well as optimised ones are presented in Fig. 4.10 for two arbitrary filling fractions 0.20 and 0.28. Accordingly, the bandgap of the 3rd gap highlighted in grey is widened and the midgap frequency is lowered at both cases. For $v_f =$ 0.20 the upper edge of bandgap is slightly increased, and the bandgap is mostly

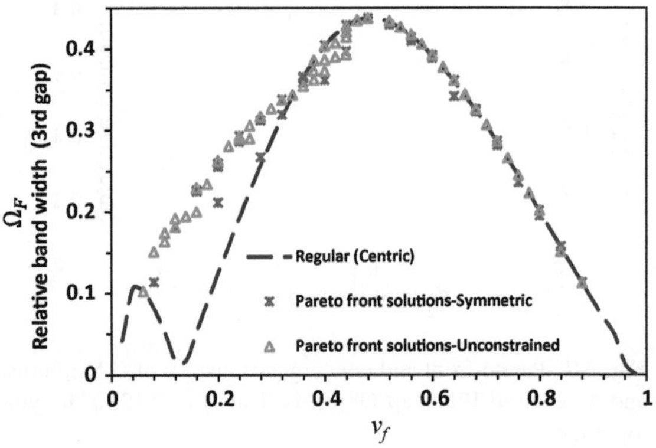

Fig. 4.8 Pareto front solutions obtained from multi-objective topology optimisation of silicon-tungsten PhP unit-cell with aspect ratio $a/h = 2$ versus filling fraction of scattering tungsten inclusion v_f, for maximised Bandgap Objective-1

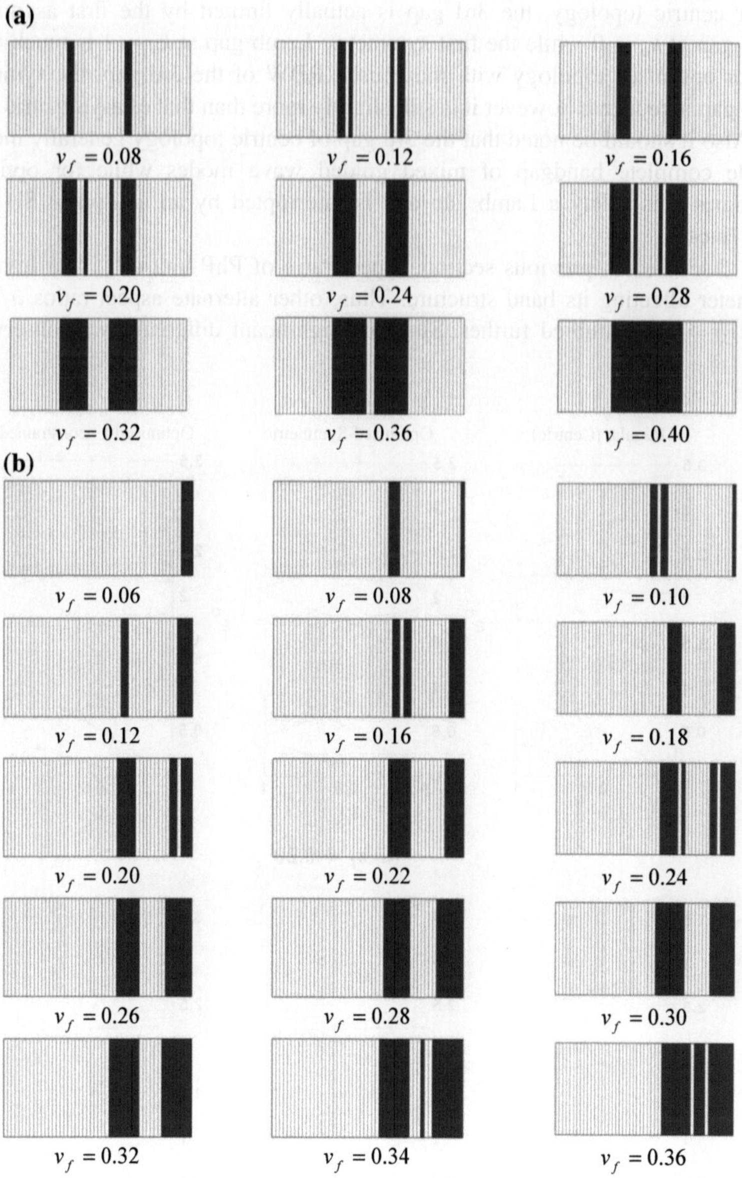

Fig. 4.9 Optimised topologies for 1D silicon-tungsten PhP unit-cell with aspect ratio $a/h = 2$ versus filling fraction of scattering tungsten inclusion v_f for maximised Bandgap Objective-1, **a** symmetric, **b** unconstrained topology

extended by its lower edge. However, for $v_f = 0.20$ the upper edge of the gap is also slightly lowered.

For centric topology, the 3rd gap is actually limited by the first asymmetric Lamb gap at $k_d = 0$ while the first symmetric Lamb gap at $k_d = 1$ is much wider. But for optimised topology with maximised RBW of the 3rd gap, the symmetric Lamb gap is reduced; however it is still slightly more than that of asymmetric Lamb gap. Also it should be noted that the 3rd gap of centric topology generally includes a wide complete bandgap of mixed guided wave modes while for optimised topologies it is solely a Lamb gap and is interrupted by an anti-plane SH mode (dash lines).

As discussed in previous section, aspect ratio of PhP unit-cell is an important parameter affecting its band structure. Thus, other alternate aspect ratios $a/h = 3$ and $a/h = 4$ are studied further. Since no significant difference was observed in

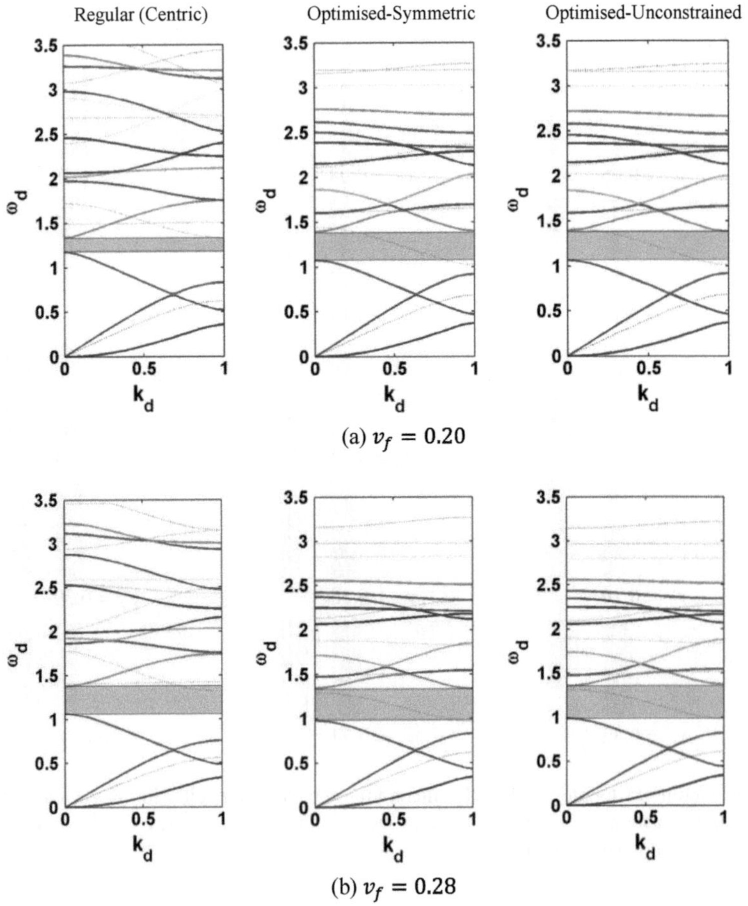

Fig. 4.10 Guided wave band structure and 3rd Lamb wave bandgap of regular PhCr with centric scattering inclusion as well as optimised symmetric-unconstrained topologies for **a** $v_f = 0.20$ and **b** $v_f = 0.28$

obtained symmetric/unconstrained topologies for aspect ratio $a/h = 2$, just optimum symmetric topologies are defined for other aspect ratios. The efficiency of achieved Pareto front solutions and optimum topologies in terms of RBW of 3rd gap are depicted in Fig. 4.11a, b for $a/h = 3$ and $a/h = 4$ respectively.

The optimised results show increased bandgap width in the range of 0.08–0.40 and 0.16–0.40 for aspect ratios 3 and 4 respectively. But for aspect ratio 3 the improved range of filling fractions is narrower and the raise in bandwidth is considerably less. Only the Pareto front solutions with improved bandgap properties, as compared to the RBW of centric topologies, are included in Fig. 4.10. Similar to the results of aspect ratio 2, the optimised results again converge to the regular centric ones for the remaining portion of filling fractions. This is observable in the relevant topologies depicted in Fig. 4.12 in where those out of ranges mentioned above are centric.

Fig. 4.11 Pareto front solutions obtained from multi-objective topology optimisation of silicon-tungsten PhP unit-cell with aspect ratio **a** $a/h = 3$ and **b** $a/h = 4$, versus filling fraction of scattering tungsten inclusion v_f, for maximised Bandgap Objective-1

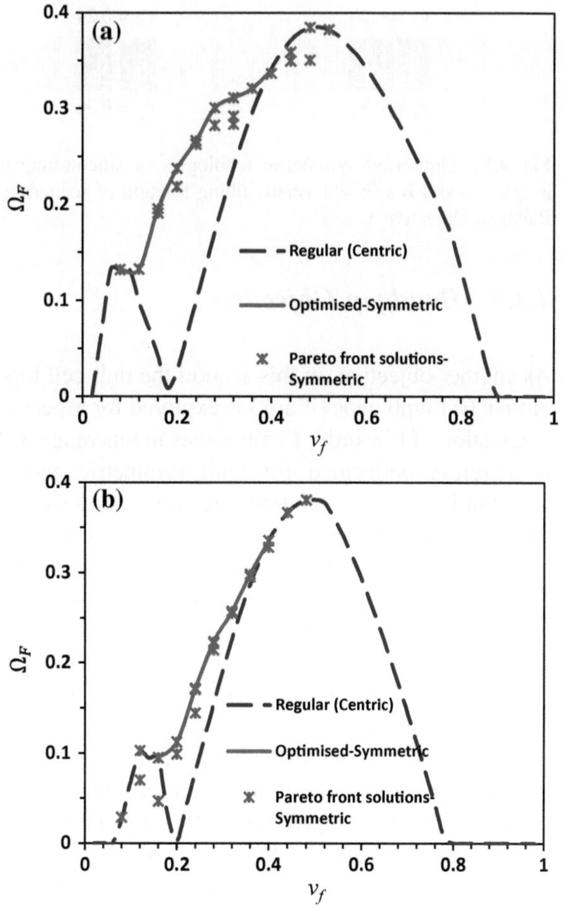

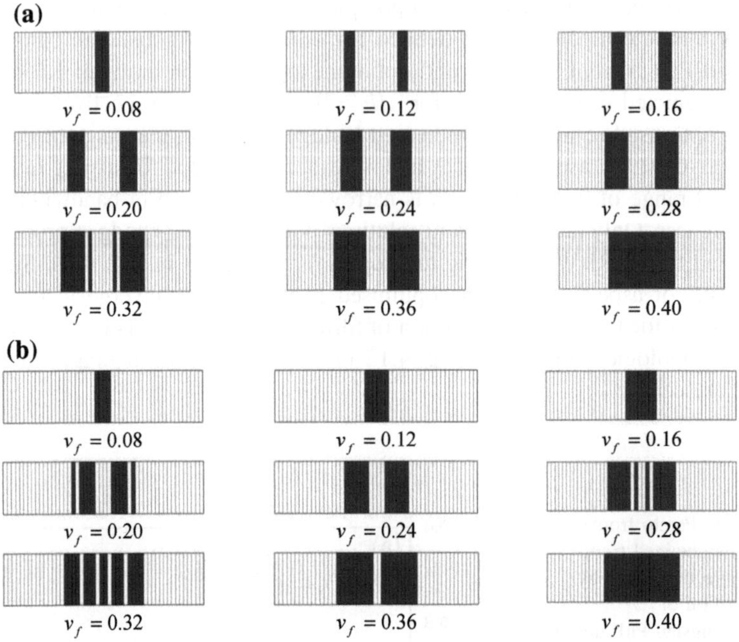

Fig. 4.12 Optimised symmetric topologies of silicon-tungsten PhP unit-cell with aspect ratio **a** $a/h = 3$ and **b** $a/h = 4$ versus filling fraction of scattering tungsten inclusion, for maximised Bandgap Objective-1

4.4.2 Bandgap Objective-2

As another objective, in this section the unit-cell topology with highest total RBW of first 10 Lamb gaps (if any) is explored for aspect ratio of 2, in order to maximise attenuation of low order Lamb waves in this range. GA multi-objective optimisation is therefore performed for both symmetric and unconstrained topologies and resultant Pareto front solutions are collected as shown in Fig. 4.14. The Pareto front and relevant convergence history for the case with reference filling fraction 0.08 is shown in Fig. 4.13. For example, Fig. 4.13a, b show optimisation of symmetric topology while Fig. 4.13c, d optimisation of unconstrained topology. The algorithm converged to the topology with desired filling fraction 0.08 (i.e. $F_2 = 0$) after 4 and 13 generations for symmetric and unconstrained optimisations respectively. Afterwards, the Pareto solution with lowest bandgap objective F_1 relates to this filling fraction which is maximised and remained unchanged for considerable number of generations (Fig. 4.13b, d).

According to Fig. 4.14, the optimised solutions have expanded the total RBW in the filling fraction range 0.08–0.60 compared to regular centric topologies. This is wider than the corresponding range of the 3rd bandgap (Fig. 4.8). More importantly, unconstrained results obtained for filling fraction range 0.08–0.20 have

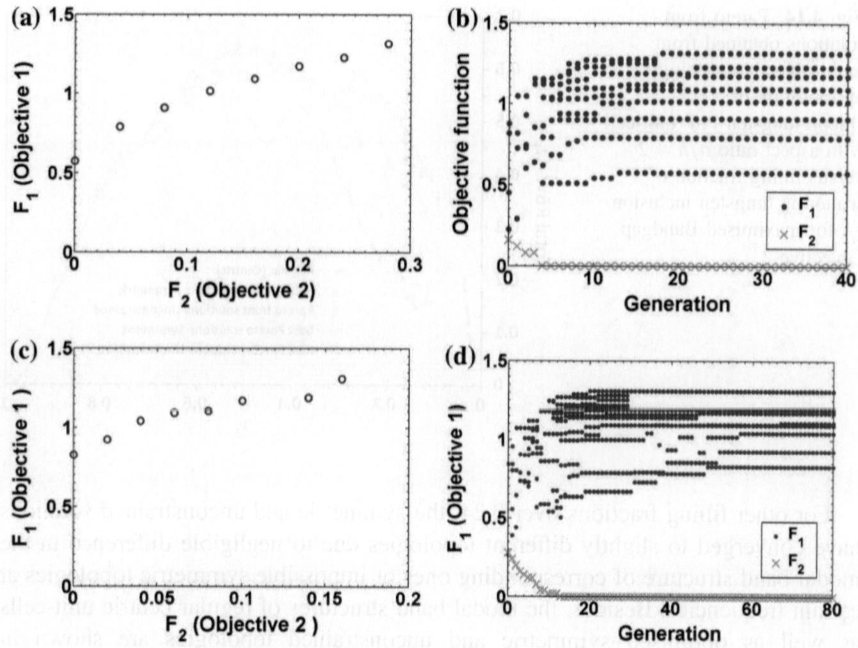

Fig. 4.13 Pareto front and convergence history of GA optimisation for PhP unit-cell with $a/h = 2$ and maximised Bandgap Objective-2 at $v_f = 0.08$, **a**, **b** symmetric topology, **c**, **d** unconstrained topology

considerably higher RBW than symmetric results with a local peak at filling fraction 0.26. For other filling fractions the symmetric and unconstrained results are close, i.e. differ negligibly. The results again converge to the centric topology for a portion of filling fractions ($v_f \geq 0.6$). However, the optimised solutions obtained for filling fraction range 0.60–0.80 have slightly lower total bandgap width as compared to those of known centric topologies. This shortcoming is arisen from the fact that the number of points searched on the Brillouin zone for any verifying iteration of GA optimisation is chosen to be 15 for the sake of computational cost. But the total bandgap width of optimised topologies and centric ones (Fig. 4.14) are calculated using 50 points which is more accurate. As shown in Fig. 4.16, the first eleven modal frequencies of interest have several sharp approaching points along the Brillouin zone, varying for different topologies. That is why the calculated total bandgap width of the first 10 likely gaps is sensitive to the chosen number of search points.

A selection of optimised topologies corresponding to the best Pareto front solutions of Fig. 4.14 are presented in Fig. 4.15a, b for symmetric and unconstrained design spaces respectively. It should be noted that the periodicity of unit-cell, the symmetric and unconstrained topologies are exactly the same for filling fractions 0.24 and 0.48.

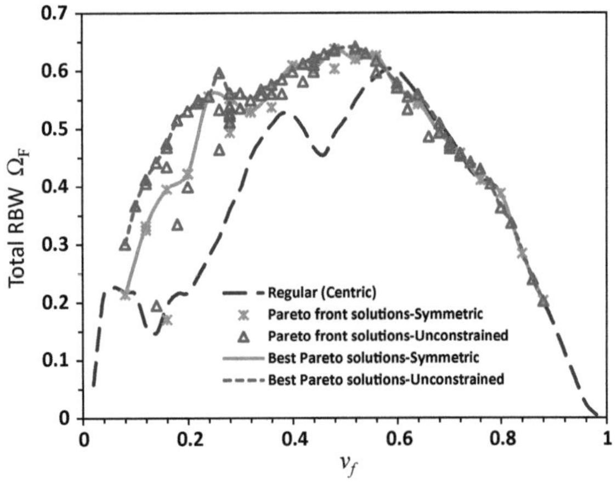

Fig. 4.14 Pareto front solutions obtained from multi-objective topology optimisation of silicon-tungsten PhP unit-cell with aspect ratio $a/h = 2$ versus filling fraction of scattering tungsten inclusion v_f, for maximised Bandgap Objective-2

For other filling fractions over 0.24, the symmetric and unconstrained solutions have converged to slightly different topologies due to negligible difference in the modal band structure of corresponding ones or impossible symmetric topologies at certain frequencies. Besides, the modal band structures of regular centric unit-cells as well as optimised symmetric and unconstrained topologies are shown in Fig. 4.14a, b for two arbitrarily chosen filling fractions 0.12 and 0.20 respectively. Lamb bandgaps opened within first eleven modes, highlighted by grey area, represent the increased total bandgap width of achieved optimum topologies as compared to regular centric ones.

For $v_f = 0.12$, the band structure associated with unconstrained topology has an additional 10th gap opened between modes 10 and 11 with respect to the symmetric one, while other 3 lower gaps have changed a bit. As for other filling fraction $v_f = 0.20$, a remarkable point is that the frequency range of first eleven modes of optimised topologies is contracted compared to the centric topology, with higher contraction in symmetric topology. This is expected as the highest total RBW (maximum bandgap widths at lowest midgap frequencies) is sought as the main objective of optimisation. In fact the dimensionless frequency range of first eleven modes is pulled down from $\omega_d = 3.38$ for centric topology to 2.65 and 2.87 for optimised symmetric and unconstrained topologies respectively. So in the same frequency range in which the eleven modal responses of centric unit-cell topology appear ($0 < \omega_d \leq 0.34$), two more modes, i.e. 12 and 13 exist for optimised topologies. Although the total bandgap width opened within the first eleven modes of unconstrained topology 0.53 is considerably higher than that of the symmetric one 0.43, it is actually effective in a wider frequency range. Nonetheless, the total bandwidth of gaps opened in the frequency range ($0 < \omega_d \leq 0.34$), taking into account extra gaps 11th and 12th, is 0.66 for unconstrained topology which is still considerably higher than that of the symmetric topology 0.54. Moreover, unlike the optimised topologies with the widest 3rd gap, the width of the first Symmetric Lamb gap is narrower than or equal to the Asymmetric one.

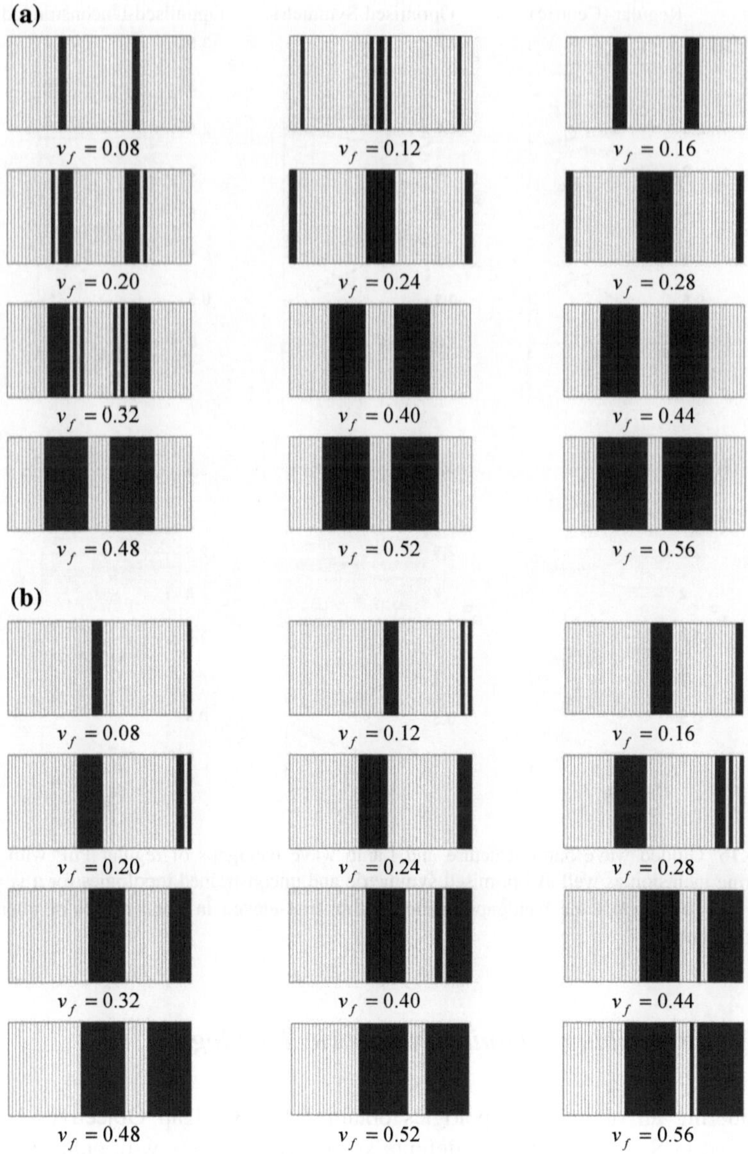

Fig. 4.15 Obtained optimised topologies of silicon-tungsten PhP unit-cell with aspect ratio $a/h = 2$ versus filling fraction of scattering tungsten inclusion v_f, for maximised Bandgap Objective-2

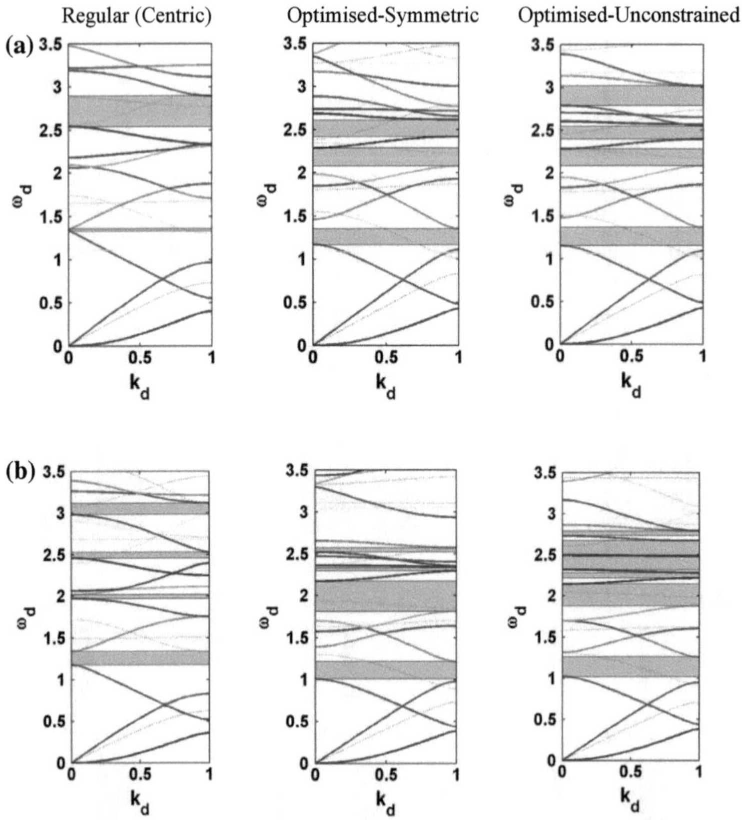

Fig. 4.16 Guided wave band structure and Lamb wave bandgaps of regular PhP with centric scattering inclusion as well as optimised symmetric and unconstrained topologies for **a** $v_f = 0.20$ and **b** $v_f = 0.48$; *Only* the bandgaps opened within first eleven in-plane modes of interest are shaded in grey

4.4.3 Prescribing Definite Symmetric Topology

Considering all optimised topologies obtained for Bandgap Objectives-1 and 2 described in Sects. 4.1 and 4.2, definite symmetric topology with enhanced bandgap efficiency is prescribed and examined herein. To serve this purpose, the symmetric topology shown in Fig. 4.17a is prescribed in which the scattering inclusion is comprised of two identical counterparts located symmetrically in the unit-cell. The only variable of this prescribed topology for specified filling fraction and aspect ratio is the relative distance of two counterparts *D* or their distance to the unit-cell's edge *b*. Assumption of unit-cell with 50 strip layers in the optimisation procedure, for the sake of computational efficiency, limited the study such that

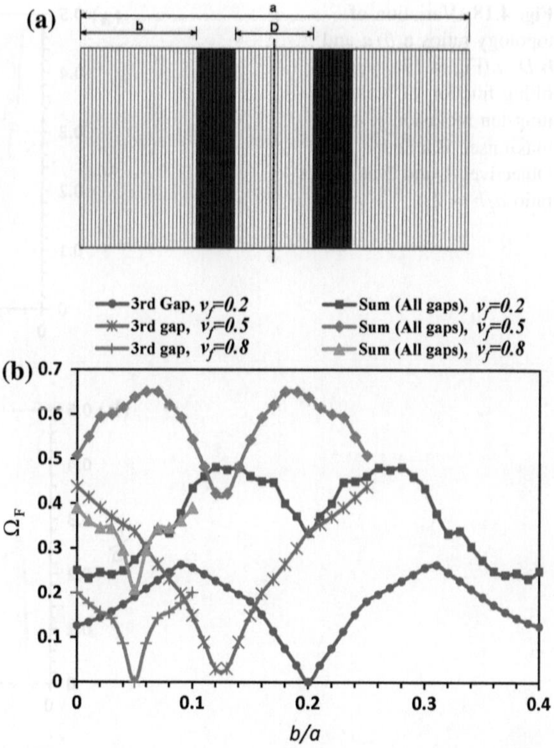

Fig. 4.17 **a** Schematic of prescribed symmetric topology for PhP unit-cell with $a/h = 2$, **b** gradient of RBW of 3rd Lamb gap and total RBW of first 10 likely gaps versus the topology ratio b/a

symmetric topology was not feasible at certain values. A more versatile unit-cell including 100 layers is therefore modelled and studied in this section.

Starting with a unit-cell of aspect ratio $r = 2$ the sensitivity of Bandgap Objectives-1 and 2 to the normalised topology variable b/a is defined at different filling fractions 0.2, 0.5 and 0.8 as depicted in Fig. 4.17b. The ratio b/a can vary from zero to a maximum value depending on the filling fraction. In Fig. 4.17b the bandgap width has two peaks at two values of b/a corresponding to two unit-cell topology modes forming the same PhP in a periodic structure. Hence, $b/a = 0$ and $b/a = Max$ stand for regular centric topology in which two counterparts stick together. For $v_f = 0.20$ the RBW of the 3rd gap is maximum at $b/a = 0.09$ and 0.31 and the total RBW of first 10 likely gaps is maximum at b/a = 0.12, 0.14, 0.26 and 0.28. Regarding $v_f = 0.50$ the maximum bandwidth of 3rd gap corresponds to the centric topology but the total RBW appears at $b/a = 0.07$ and 0.18. Finally the best prescribed topology for filling fraction $v_f = 0.80$ to gain both Bandgap Objectives is the centric one.

For simplicity the focus of study is narrowed further to the best unit-cell's topology mode with higher b/a (right side peaks in Fig. 4.17b) or least D/a corresponding to unit-cell with concentrated scattering inclusion at the centre. Then all feasible symmetric filling fractions from 0.02 to 0.9 are analysed to define the

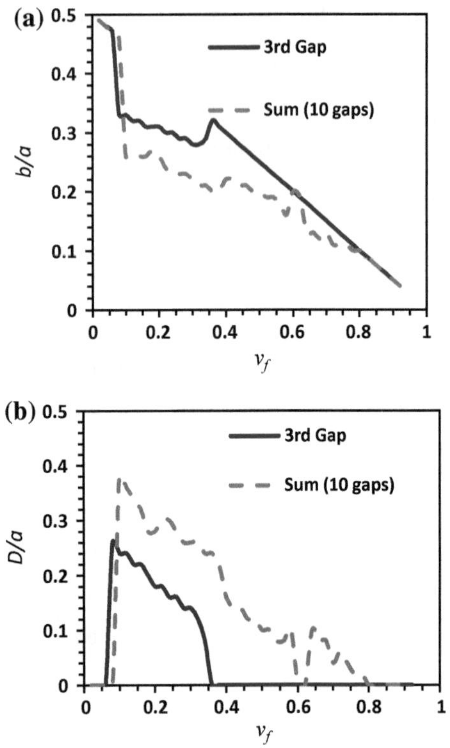

Fig. 4.18 Variation of topology ratios **a** b/a and **b** D/a (Fig. 4.17a) versus filling fraction of scattering tungsten inclusion v_f for maximised Bandgap Objectives-1 and 2 at aspect ratio $a/h = 2$

topology variable b/a or D/a associated with the unit-cell having maximum RBW as shown in Fig. 4.18. According to these results, the best topology variable D/a has steeper variation versus filling fraction as compared to b/a. It means that the RBW is essentially determined by the external edges of scattering inclusion rather than its internal edges. The efficiency of the best prescribed topology versus the filling fraction can now be investigated relative to the centric topology and previously optimised topologies. So, the relevant results are synthesized for this purpose in Fig. 4.19a, b showing RBW of 3rd gap and total RBW of first 10 likely gaps respectively.

From Fig. 4.19a it is evident that prescribed topologies are in excellent agreement with optimised ones in terms of RBW of the 3rd gap except of unconstrained topologies which have slightly wider gap width. In Fig. 4.18a the ratio b/a varies from 0.28 to 0.33 for the filling fraction range of 0.06–0.40 in which the bandwidth of the 3rd gap is raised relative to the centric topology. Thus, over this filling fraction range, the RBW of third gap corresponding to unit-cell topologies with constant variable $b/a = 0.30$ and $b/a = 0.31$ are also included in Fig. 4.19a.

The topologies with constant b/a have negligible deviation from the best topologies. Better agreement is observed at higher filling fractions for $b/a = 0.30$ and at lower filling fractions for $b/a = 0.31$. Figure 4.19b also shows good

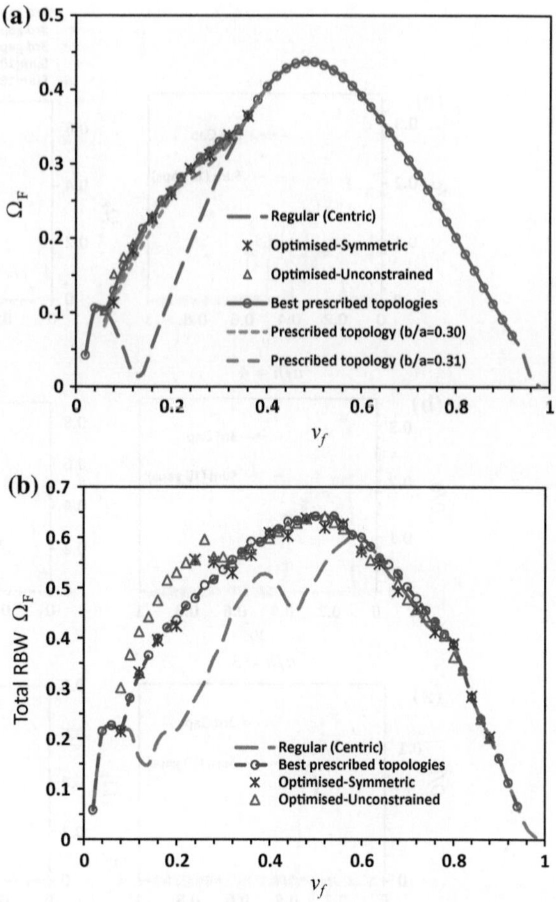

Fig. 4.19 RBW of best prescribed topologies as well as centric and optimised topologies of silicon-tungsten PhP unit-cell with aspect ratio $a/h = 2$ versus filling fraction of scattering tungsten inclusion v_f, for maximised **a** Bandgap Objective-1 and **b** Bandgap Objective-2

agreement of the best prescribed topology and the optimised symmetric topology in terms of total RBW of the first 10 likely gaps. However, optimised symmetric topologies at filling fractions 0.24 and 0.28 and optimised unconstrained topologies in filling fraction range 0.08–0.32 are considerably better than prescribed topologies. Also, the prescribed topologies show a minor improved total bandwidth in the range of 0.64–0.80.

Furthermore, unit-cell aspect ratios of 1, 3 and 4 are studied in order to investigate the efficiency of prescribed symmetric topologies at other aspect ratios. The relevant results depicted in Fig. 4.20 show that for $r = 1$ the best prescribed topology for the widest 3rd gap is the centric one ($D/a = 0$) at all filling fractions. But the total RBW is widened only in filling fraction range of 0.18–0.36 relative to the centric topology. For the aspect ratios 3 and 4 again the best prescribed topologies exactly coincide with the optimised symmetric topologies. The total bandwidth of the first 10 likely gaps is also enlarged at both aspect ratios over a

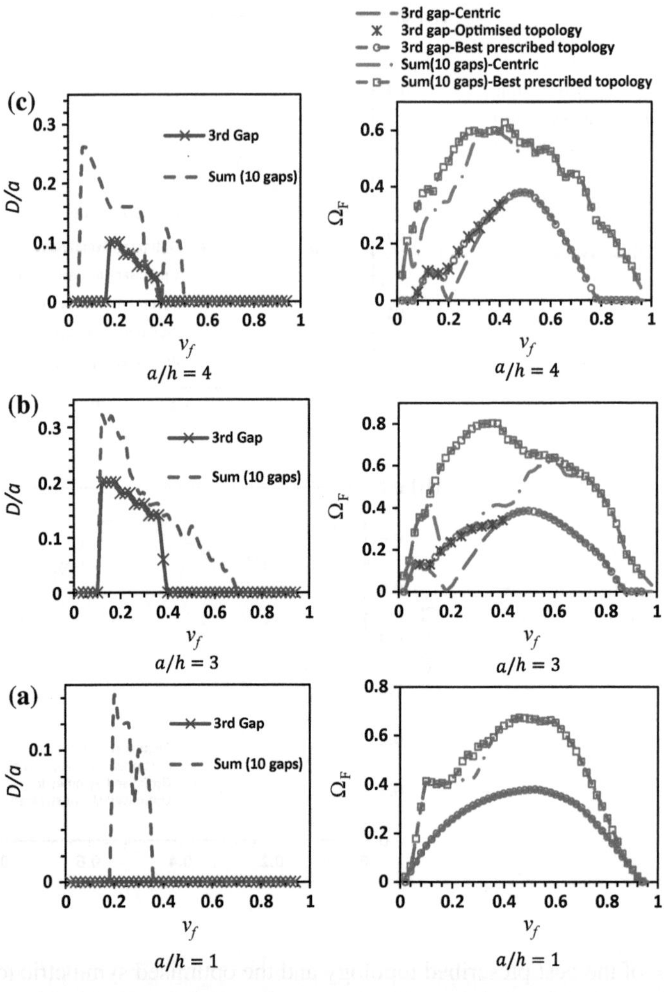

Fig. 4.20 Left: Variation of topology ratio D/a (Fig. 4.17a) and Right: RBW of corresponding prescribed topologies as well as centric and optimised topologies versus filling fraction of scattering tungsten inclusion v_f for maximised Bandgap Objectives-1 and 2 at different aspect ratios, **a** $a/h = 1$, **b** $a/h = 3$ and **c** $a/h = 4$

significant range of filling fractions 0.10–0.72 and 0.04–0.50 for aspect ratios 3 and 4 respectively.

Consequently, the prescribed topologies introduced in this section are definite reliable substitutes for optimised topologies in a wide range of filling fractions especially in terms of RBW of the 3rd gap. Getting optimised topologies of any arbitrary filling fraction through genetic algorithm demands implementation of unit-cell having huge number of strip layers leading to an enormous design space. Hence the simple definite topologies prescribed herein can be reliably used for example in a continuous variation of filling fraction through a graded structure with nearly maximised bandgap efficiency.

4.5 Frequency Response Analysis of Finite PhP Structures

The modal band structure of 1D PhP unit-cell with infinite periodicity was defined in previous sections by applying Bloch-Floquet periodic boundary condition, and topologies with optimised bandgap properties were investigated for various filling fractions. Now, this section is dedicated to study the frequency response and elastic wave transmittance of finite plate structures comprised of consecutive silicon-tungsten PhP unit-cells. Different arrangements of topologies are studied to evaluate the depth and width of Lamb bandgaps associated with uniform and graded PhP structures with regard to the band structure of perfectly periodic unit-cells. In other words, these cases are selected so that comparing their transmission spectrums demonstrates the effect of unit-cell optimisation and unit-cell gradient throughout the structure on its bandgap properties. Besides, the credibility of achieved modal band structure of associated PhP unit-cells is assessed.

For this purpose, initially 6 cases are introduced as displayed in Fig. 4.21 to mainly focus on their 3rd Lamb gap (Bandgap Objective-1). Then cases 7–9

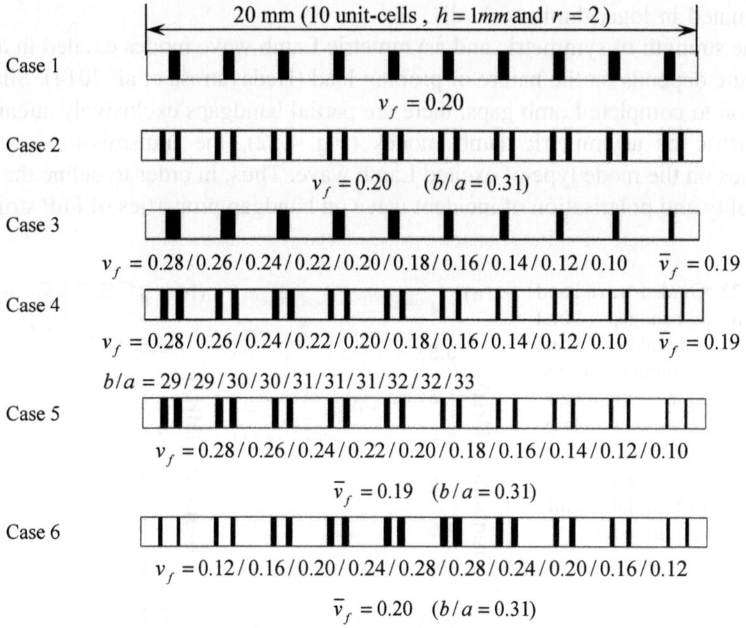

Fig. 4.21 Finite PhP structure cases including 10 unit-cells with aspect ratio $a/h = 2$ and thickness $h = 1\,\mathrm{mm}$ (optimised and prescribed topologies correspond to maximised Bandgap Objective-1)

presented in Fig. 4.27 are defined to investigate their total bandgap width within first few modes (Bandgap Objective-2). All cases are plate structures including 10 unit-cells of aspect ratio $a/h = 2$ and thickness $h = 1$ mm leading to a total length of 20 mm. The average filling fraction of cases is set to 0.2 or a near value 0.19, from the filling fraction range in which significant improvement was observed in bandgap efficiency of optimised unit-cells.

Cases 1 and 2 are uniform PhP structures with centric and best prescribed topology respectively. Prior to frequency response analysis of these cases, in order to get vision about their expected bandgap properties, the modal band structure of relevant unit-cells are defined for filling fraction 0.2 and thickness $h = 1$ mm, as shown in Fig. 4.22. The frequency range 0–4.2 MHz covering the first eleven in-plane modes is chosen for both topologies. The Lamb gaps opened within this frequency range are marked by grey area. Also, the symmetric Lamb modes are indicated by bold dots to identify the exclusive bandgaps of Symmetric and Asymmetric Lamb modes.

For frequency response analysis of PhP structure, its FEM model is generated by standard FEM solver ANSYS APDL (*ANSYS® Academic Research, Release 14.5*) and the reliability of results obtained from developed model are validated. Any unit-cell is modelled with 100 longitudinal and 25 transversal elements using *ANSYS PLANE183* quadratic 8 node elements. A harmonic probing load of amplitude 0.1 N is applied to the left edge over the frequency range 0–4.2 MHz and the relative transmission spectrum of induced oscillation to the top-right corner is determined in logarithmic scale dB.

The strength of symmetric and asymmetric Lamb wave modes excited in a plate structure depends on the nature of probing load (Hedayatrasa et al. 2014). Since, in addition to complete Lamb gaps, there are partial bandgaps exclusively attenuating symmetric or asymmetric lamb modes (Fig. 4.22), the transmission spectrum depends on the mode type of excited Lamb wave. Thus, in order to define the effect of quality and polarisation of incident wave on bandgap properties of PhP structure,

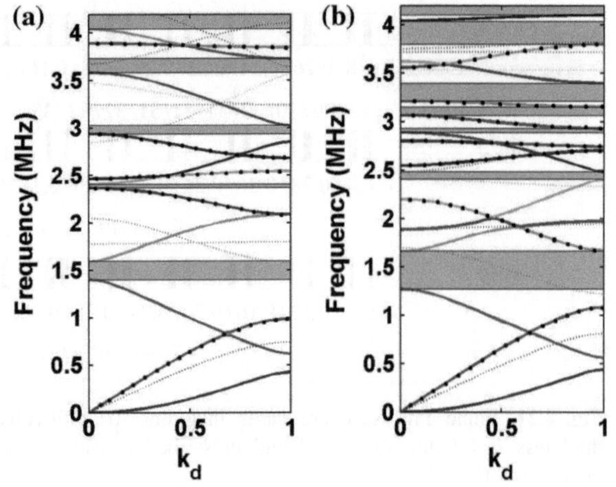

Fig. 4.22 Guided wave band structure and bandgaps of first 11 Lamb modes along the first irreducible Brillouin zone for unit-cell with $h = 1$ mm and $v_f = 0.20$, **a** centric topology and **b** best prescribed topology for Bandgap Objective-1 (bold dots signify Symmetric Lamb mode)

Table 4.3 Loading cases applied for frequency response analysis

No.	Polarisation	Amp. (N)	Description
1	Transversal	0.1	Point load on top left-corner, no bounded silicon unit-cell
2	Transversal	0.1	Point load on top left-corner, with bounded silicon unit-cell
3	Longitudinal	0.1	Point load on top left-corner, with bounded silicon unit-cell
4	Longitudinal	0.1	Distributed load on left edge, with bounded silicon unit-cell

the transmittance of case structure 1 is defined for 4 different loading cases as listed in Table 4.3.

Loading Case 1 corresponds to the model with a point load of transversal polarisation applied to the top-left corner of structure. Of course, point load in this 2D cross sectional model is equivalent to a uniform load per unit width of plate structure along z-axis. The load's amplitude is selected to be small enough in order to avoid large displacements and so geometrical nonlinearity. Loading Case 2 is similar to the loading Case 1 with the only difference that a pure silicon unit-cell is added at both ends between the probing point and structure as well as the receiving point and structure (Fig. 4.23).

The transmission spectrums of Case 1 subjected to all 4 loading cases is presented in Fig. 4.22a. The dips marked by numbered arrows stand for attenuation of frequency response due to existence of bandgaps. The deeper a gap is the higher attenuation occurs. The results show improved bandgap properties for loading Case 2 compared to loading Case 1, as deeper and wider gaps appear in this case, showing better agreement with obtained modal band structure. This is evident specially in dip 1 concerning the 3rd Lamb gap. Although the loading case 1 can reliably provide the frequency response of the finite PhP structure of interest, the loading case 2 shows better agreement with obtained modal band structure of unit-cell with infinite periodicity. In fact the bounded pure silicon unit-cells make better simulation of guided waves for the incident oscillation approaching the PhP structure and transmitted oscillation leaving it. In other words, the lack of periodicity in the boundary PhP cells is somehow compensated by uniform silicon extension (instead of a free boundary) making better agreement with band structure premised on perfect periodicity. So in all other loading cases this condition is applied where a pure silicon unit-cell is added to both ends of structure.

Fig. 4.23 Bounding a pure silicon unit-cell two both ends of structure for uniform Lamb wave incident and improved bandgap efficiency

Regarding the spectrum of loading Case 2, dips 1, 3 and 4 correspond to the 3rd, 8th and 9th Lamb gaps around 1.50, 2.90 and 3.60 MHz as shown in relevant band structure Fig. 4.22a. Dips 5 and 6 are related to the exclusive bandgaps of asymmetric Lamb waves. Moreover the lower band of dip 3 is extended due to existence of an asymmetric Lamb gap beside the 8th Lamb gap. The symmetric Lamb modes are indicated by bold dots in modal band structure Fig. 4.22 to highlight the exclusive bandgaps of symmetric and asymmetric Lamb modes. Although dip 2 corresponding to the 5th Lamb gap is expected to happen around 2.40 MHz, it has been faded by strong adjacent asymmetric Lamb gap 6. It is well defined that applying a transversal load on just top side of plate structure like loading Case 2 causes dominant excitation of asymmetric Lamb waves with leading transversal oscillations. That is why strong attenuation of asymmetric Lamb gaps is observed for this loading case.

In loading Case 3 a probing load of longitudinal polarisation is applied to the top left corner and transmission of longitudinal oscillation is defined. Loading Case 3 still has asymmetric nature; nonetheless its horizontal polarisation boosts symmetric Lamb modes. In the spectrum of loading Case 3, as compared to the loading Case 2, the asymmetric Lamb gaps 5 and 6 disappear and Lamb gap 2 becomes recognisable at around 2.4 MHz. Regarding loading Case 4 the left edge of plate structure is subjected to a uniformly distributed longitudinal load in order to excite pure symmetric Lamb modes. The relevant spectrum clearly shows appearance of robust symmetric Lamb gaps 7, 8, 9 and 10 around 1.50, 2.60, 3.50 and 4.00 MHz respectively. The symmetric Lamb gap 7 is developed from just below 1.0 MHz to around 2.1 MHz, and the symmetric Lamb gap 9 from just below 3.00 MHz to around 3.90 MHz as predicted by related modal band structure Fig. 4.22a. Plus, the results show that the depth of wave attenuation changes significantly for different loading cases as they lead to Lamb waves with various constructions. For instance, the Lamb gap 1 (loading Case 2) and symmetric Lamb gap 7 (loading Case 4) show very different attenuation depths of 37.3 and 105.0 dB at 1.50 MHz which is the midgap frequency of 3rd Lamb gap.

The transmission spectrums of all 6 cases introduced in Fig. 4.21 subjected to loading Case 2 are then collected in Fig. 4.24b. Comparing the frequency response of introduced structures enables the assessment of their bandgap efficiency in terms of the 3rd Lamb gap (Bandgap Objective-1).

The Case 2 is actually a uniform PhP structure with filling fraction $v_f = 0.20$ like Case 1 but with unit-cells of best prescribed topology for maximum bandwidth of the 3rd gap. As shown in the previous section, the best prescribed topology of unit-cell with $v_f = 0.20$ and $a/h = 2$ is in excellent agreement with optimised topology. Cases 3–5 are alternative graded PhP structures (type A) in which the filling fraction of unit-cell evolves gradually from 0.1 to 0.28 providing average filling fraction of 0.19 close to introduced uniform structures with $v_f = 0.20$. But different topologies are considered for unit-cells of these cases as illustrated in Fig. 4.21. Finally the Case 6 with another alternative graded pattern (type B) of average filling fraction $v_f = 0.20$ is introduced by unit-cells having the best

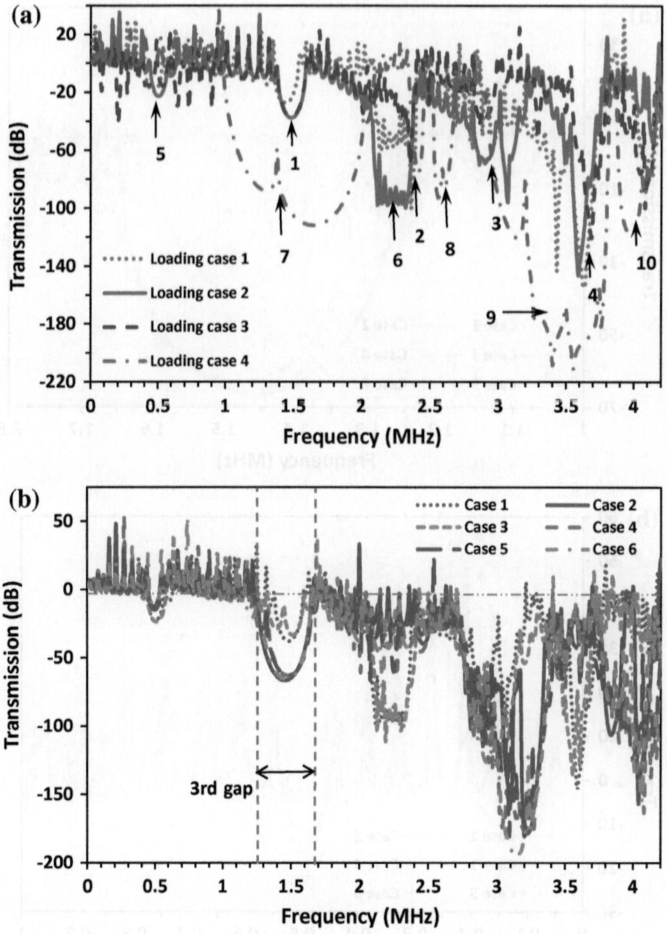

Fig. 4.24 Frequency response of PhP structure cases (Fig. 4.21) over the frequency range 0–4.2 MHz covering first 11 in-plane modes, **a** Case 1 subjected to various loading cases (Table 4.3), and **b** cases 1–6 subjected to loading Case 2

prescribed topology. From Fig. 4.24b it is easily understood that using unit-cell with best prescribed topology makes the 3rd Lamb gap significantly wider and deeper.

However, the first exclusively asymmetric Lamb gap around 0.50 MHz is degraded. The 3rd bandgap of Case 2 with the best prescribed topology is extended from 1.28 to 1.66 MHz as shown in Fig. 4.22b. This is considerably wider than that of Case 1 with centric topology extended from 1.4 to 1.59 MHz. For clarity the two sections of transmission spectrum in Fig. 4.24b in the low frequency range as well as the frequency range corresponding to the 3rd gap are depicted in Fig. 4.25a, b

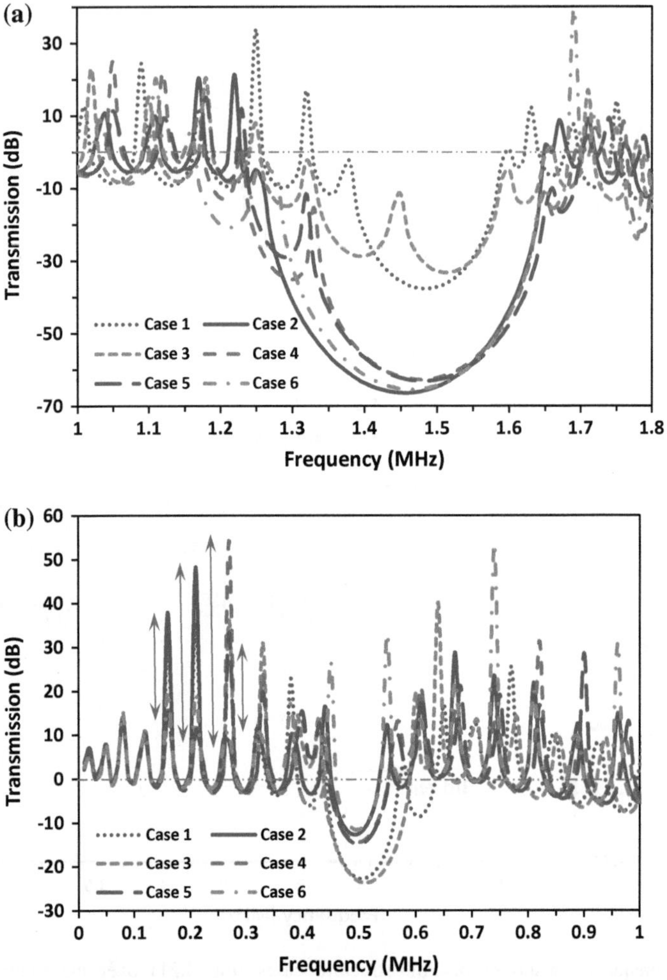

Fig. 4.25 Frequency response of PhP structure cases (Fig. 4.21) over the frequency range **a** 1.00–1.80 MHz highlighting the differences in the 3rd Lamb bandgap width and depth, and **b** 0–1.00 MHz highlighting the differences in low frequency response

respectively. Accordingly, the attenuation depth of the 3rd gap is increased from 37.3 (for centric topology) to 66.4 dB (for the best prescribed topology).

Moreover, the results confirm that applying different arrangements of unit-cells with various filling fractions to introduce graded PhP, with almost the same average filling fraction, nearly maintains its bandgap efficiency. Using a graded pattern of centric unit-cells (Case 3) decreases the depth of the 3rd gap from 37.3 to 33.3 dB with slightly lower filling fraction of 0.19 compared to that of Case 1.

Likewise, the graded Case 4, 5 and 6 have the 3rd gap depth of order 62.8, 63.2 and 65.5 dB with slight variation with respect to that of the corresponding uniform structure (Case 2) 66.4 dB. Case 5 with an arrangement of the best prescribed topologies has slightly less gap depth compared to Case 6 with uniform prescribed topology of $b/a = 0.31$. But, Case 5 shows higher attenuation in low frequency band of the 3rd gap as expected from Fig. 4.19a. Although the graded structure Case 6 has nearest gap depth to the corresponding uniform structure Case 2, it has filling fraction of 0.2 exactly the same as Case 2. However the Case 6 with gradient pattern type B shows wider 3rd bandgap as compared to Case 4 and 5 with gradient pattern type A and even corresponding uniform Case 2.

Of course, gradient of filling fraction of unit-cells throughout the PhP structure changes its overall stiffness and consequently principal modal response. Scrutinizing the few first modal responses of graded structures with corresponding uniform ones as shown in Fig. 4.25b confirms this argument.

For example, the frequency response at the 5th–8th modal frequencies of PhP structures is highly affected by the gradient of unit-cell. The transmission at the 5th modal frequency is highly attenuated by 23 dB form Case 2 to Case 4 or 5, by 18 dB from Case 2 to Case 6 and by 16 dB from Case 1 to Case 3. The same trend is observed for the 6th frequency mode with even higher reduction rate. In contrast, the frequency response of the 7th mode is escalated by 45 dB from Case 2 to Case 4, by 24 dB from Case 2 to Case 5 and by 27 dB from Case 1 to Case 3. A transmission rise is also present for 8th modal response. But, gradient pattern type B (case 6) shows negligible influence on transmission spectrum at the 7th and 8th modal frequencies.

Spatial gradient of steady state frequency response at bandgap frequency 1.5 MHz along the cross section of the two selected PhP structure Case 2 and 5 is presented in Fig. 4.26 in terms of dB for loading Case 2. Contours of longitudinal and transversal displacement components are shown separately. The exponential decay of both polarisations through consecutive PhP unit-cells is evident in both cases. Through thickness profiles of oscillations with very low longitudinal displacement at the mid-plane and almost uniform transversal displacement apparently show the dominant excitation of asymmetric Lamb mode under loading Case 2.

As for second set of cases aimed at maximising the Bandgap Objective-2, three structures are assumed. Case 7 and 8 are uniform PhP structures of filling fraction 0.2 with best prescribed topology and optimised unconstrained topologies respectively. As shown earlier in Fig. 4.19b these two topologies have considerable different efficiencies in total RBW of the first 10 Lamb gaps. Last Case 9 is a graded PhP of type A with the best prescribed topology and average filling fraction of 0.19 (Fig. 4.27).

The modal band structure of unit-cells included in Case 7 and 8 as well as centric one are determined for $a/h = 2$ and $h = 1$ mm as shown in Fig. 4.28. Again the frequency range 0–4.2 MHz, including the first eleven Lamb modes of interest for all 3 cases, is selected for analysis. Then the transmission spectrums of these cases subjected to loading Case 2 are calculated as collected together in Fig. 4.29. The spectrums of Case 1 and 2 are also included in Fig. 4.29a as a reference to assess the effect of optimisation on total band width of plate structures.

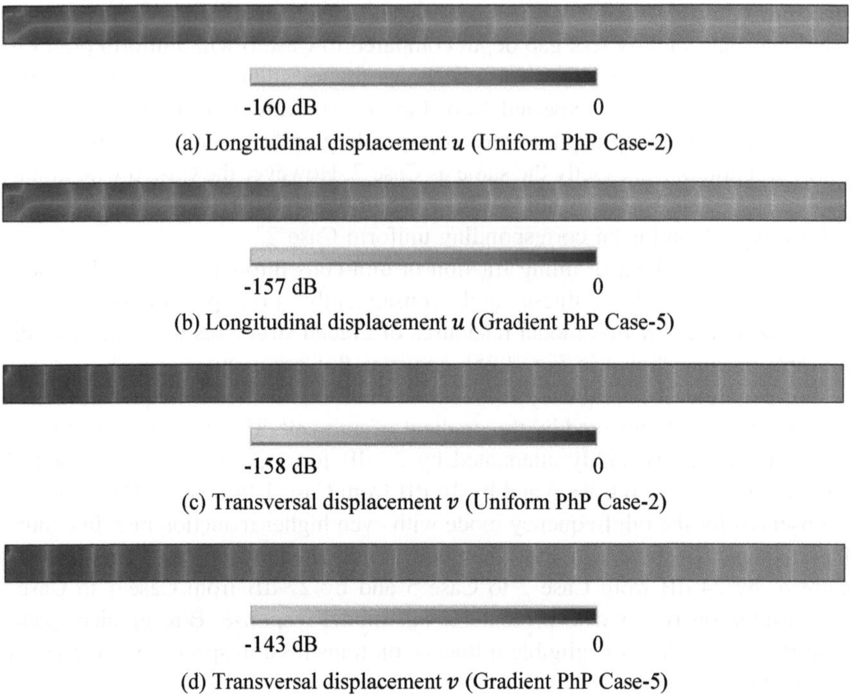

Fig. 4.26 Contour of normalised displacement components in logarithmic scale dB at bandgap frequency 1.5 MHz for **a**, **c** optimised uniform PhP Case 2, and **b**, **d** optimised gradient PhP structure case 5 (Fig. 4.21) subjected to loading case 2

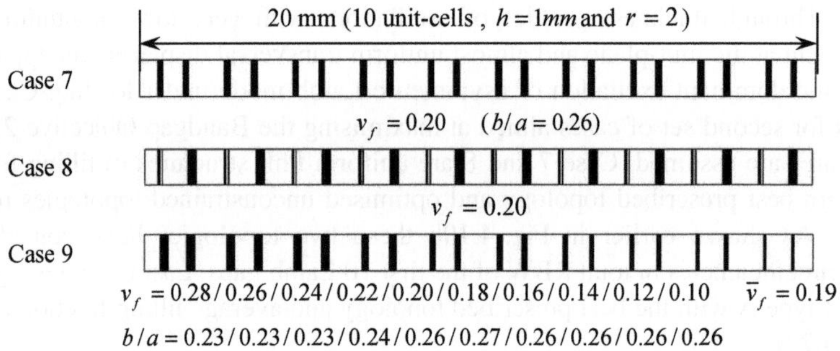

Fig. 4.27 Finite PhP structure cases including 10 unit-cells with aspect ratio $a/h = 2$ and thickness $h = 1$ mm (optimised and prescribed topologies correspond to maximised Bandgap Objective-2)

From Fig. 4.29a the improved wide bandgap efficiency of Case 7 with prescribed symmetric topology is obvious as compared to Case 1 with centric topology. Case 8 with optimised unconstrained topology shows even higher relative

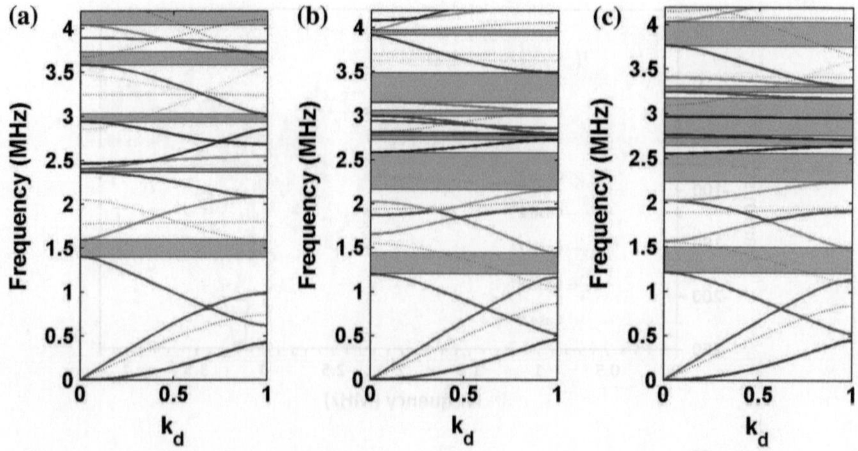

Fig. 4.28 Guided wave band structure and bandgaps of first 11 Lamb modes along the first irreducible Brillouin zone for unit-cell with $h = 1$ mm and $v_f = 0.2$, **a** centric topology, **b** best prescribed topology and **c** optimised unconstrained topology for Bandgap Objective-2

bandgap efficiency as compared to Case 7. A narrow dip is present just below 2 MHz related to partial Lamb gap existing between two subsequent asymmetric Lamb modes in this range (Fig. 4.28c). Deeper low frequency attenuation is observed in the range 2.00–3.50 MHz leading to higher relative band width of Case 8 with respect to Case 7. As predicted by the relevant band structure (Fig. 4.28) a set of bandgaps are present in the ranges 2.1–3.50 MHz and 2.20–3.30 MHz for Cases 7 and 8 respectively, however with lower attenuation in the range 2.60–3.20 MHz for Case 7.

By comparing the transmission spectrum of Case 7 and 2 one question arises: why does Case 7 provide wider and deeper attenuation for the 3rd Lamb gap around 1.5 MHz while this is expected of Case 2. This observation is justified by the fact that the dip shown around 1.50 MHz for Case 7 is concerned with asymmetric Lamb gap. According to Fig. 4.28b the asymmetric Lamb gap on the left side of the 3rd Lamb gap from 1.20 to 1.65 MHz is larger than the symmetric one on the right side from 1.16 to 1.44 MHz (unlike Fig. 4.22b). So, as mentioned earlier, applied loading Case 2 excites asymmetric Lamb waves dominantly leading to a wide and deep asymmetric Lamb gap in frequency response of Case 7.

The graded structure of Case 9 with average filling fraction 0.19 exhibits good bandgap efficiency with respect to its equivalent uniform structure Case 7 with filling fraction 0.2. As for Case 9 a bit lower attenuation is observed for the 3rd Lamb gap around 1.5 MHz and high frequency range 3.00–3.50 MHz, while showing somewhat deeper attenuation in the range of 1.70–3.00 MHz. This relative inconsistency could also be explained by their different average filling fractions.

Again the low frequency section depicted in Fig. 4.29b shows how the principal modal responses of PhP are altered by introducing gradient structure with varying unit-cells. For example, the transmission is highly diminished at modal frequencies

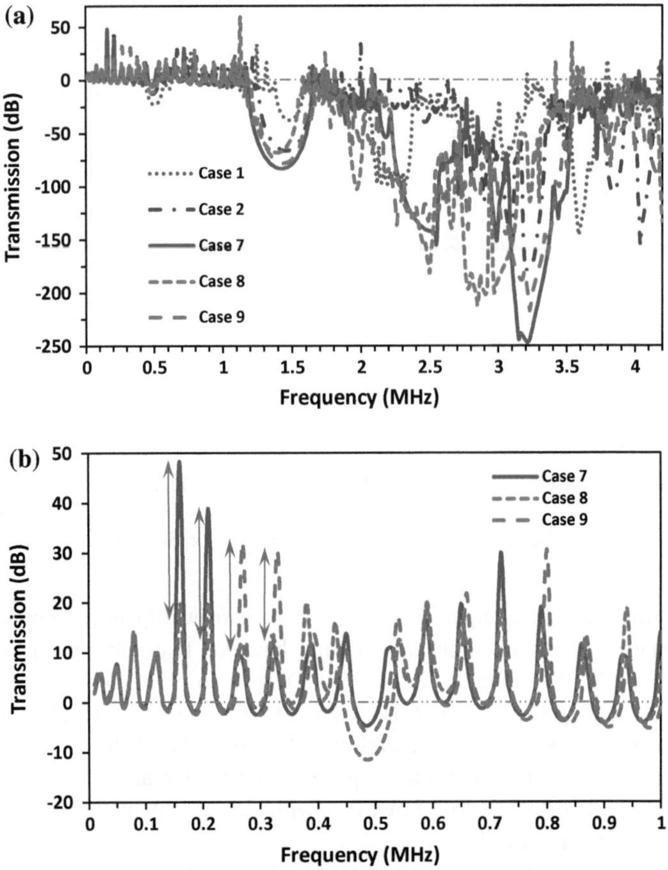

Fig. 4.29 Frequency response of PhP structure cases (Figs. 4.21 and 4.27) over the frequency range **a** 0–4.20 MHz covering first 11 in-plane modes, and **b** 0–1.00 MHz highlighting the differences in low frequency response

5th and 6th by in order 32 and 27 dB, and highly escalated at subsequent modal frequencies 7th and 8th by in order 22 and 18 dB respectively. Hence, the low frequency response of this graded PhP is considerably manipulated while maintaining its high frequency bandgap properties. In fact multiscale functionality is observed for this structure: in unit-cell scale of length 2 mm the bandgap properties are altered from around 1.50 MHz, while in structure scale of length 20 mm the basic modal responses are influenced at proportionally lower frequencies of around 0.15 MHz.

4.6 Concluding Remarks

Basic study was performed on topology optimisation of 1D bi-material PhP unit-cell for highest specific Lamb bandgap width at prescribed filling fraction of tungsten scattering inclusion. Moreover the bandgap efficiency and multiscale functionality of gradient PhP structures comprising of unit-cells with various filling fractions and optimised topologies were investigated. Silicon and tungsten were used in this research as the constitutive materials of bi-material PhP. However, the bandgap efficiency essentially depends on the contrast in density and elastic properties of constitutive materials. So for other relative properties new optimisation should be performed.

Specialised FEM model was developed for this purpose calculating the dispersion curves and modal band structure of guided waves. The filling fraction dependency of low order Lamb gaps for regular topology of unit-cell with centric scattering layer was studied considering various aspect ratios. Topology optimisation was then performed successfully through NSGA-II in order to maximise the RBW of low order Lamb waves around particular filling fraction of scattering tungsten inclusion.

Optimised topology was defined for two different bandgap objectives separately. Bandgap Objective-1 was defined as the widest RBW of lowest Lamb gap while Bandgap Objective-2 was proposed as the widest total RBW of 10 likely gaps within the first 11 in-plane modes of interest. Besides, two types of design space with symmetric topology as well as constraint free topology were examined. Simplified symmetric topology was then prescribed inspired by obtained optimum ones to somehow introduce definite filling fraction based topology with improved bandgap efficiency which is relatively easy to fabricate.

Finally, the frequency response analysis of a few cases confirmed the enhanced bandgap efficiency of finite PhP structures comprising of consecutive unit-cells of achieved topologies. Vibration attenuation occurred at the same frequency ranges predicted by modal band analysis of unit-cell. Depending on the excited Lamb wave mode type (i.e. symmetric or asymmetric) high attenuation was observed coinciding with symmetric, asymmetric and complete Lamb gaps.

The following specific conclusions were drawn from the results:

- The bandgap properties of optimised topologies were assessed as compared with those of regular centric topology as a consequence of which the optimised topologies exhibited better bandgap efficiency in specific filling fraction range for both bandgap objectives. The optimised topology converged to centric topology for the remaining portion of filling fractions.
- The Bandgap Objective-2 increased over considerably wider range of filling fractions compared with that of Bandgap Objective-1. The symmetric constraint on topology of unit-cell had no considerable effect on the efficiency of obtained topologies for Bandgap Objective-1. But major increase was observed in bandgap efficiency of unconstrained topologies, with respect to symmetric ones, achieved for Bandgap Objective-2 at low filling fraction range.

- Studying the modal band structures revealed that the existence of first complete bandgap of mixed wave modes (including anti-plane SH modes) is penalised by widening Lamb gap through optimisation.
- Simplified prescribed symmetric topology, inspired by the optimised ones, proved to be definite reliable substitutes for optimised topologies in a wide range of filling fractions particularly for Bandgap Objective-1.
- Harmonic analysis of representative finite structures confirmed that introducing a gradient PhP, with almost the same average filling fraction as uniform PhP, closely maintains its bandgap efficiency. It is also shown how the basic modal response of PhP structure is altered due to different overall structural properties.
- Such multiscale lattice structure could have multi-functionality at two scales. On the one hand, the micro-mechanical topology is optimised in unit-cell scale for maximised bandgap efficiency. On the other hand, the filling fraction of unit-cell varies to create macro-mechanically graded structure and serve structure scale objectives, e.g. subwavelength manipulating of wave or satisfying desired structural weight and/or stiffness efficiency.

References

Chen, J. -J., Zhang, K. -W., Gao, J., & Cheng, J. –C. (2006). Stopbands for lower-order Lamb waves in one-dimensional composite thin plates. *Physical Review B*, *73*(9), 094307.

Chen, J. –J., & Han, X. (2010). The propagation of Lamb waves in one-dimensional phononic crystal plates bordered with symmetric uniform layers. *Physics Letters A*, *374*(31–32), 3243–3246.

Degertekin, F. L., Honein, B., & Khuri-Yakub, B. (1995). Efficient excitation and detection of Lamb waves for process monitoring and NDE. In *1995 IEEE Ultrasonics Symposium Proceedings* (pp. 787–790).

Felippa, C. A. (2004). Multi freedom constraints. *Introduction to finite element methods*. Boulder, Colorado, 80309-0429, USA: University of Colorado.

Gao, J., Zou, X.-Y., Cheng, J.-C., & Li, B. (2008). Band gaps of lower-order Lamb wave in thin plate with one-dimensional phononic crystal layer: Effect of substrate. *Applied Physics Letters, 92*(2), 023510.

Gopalakrishnan, S., Chakraborty, A., & Mahapatra, D. R. (2007). *Spectral finite element method: Wave propagation, diagnostics and control in anisotropic and inhomogeneous structures*. Springer. Doi: https://doi.org/10.1007/978-1-84628-356-7

Hedayatrasa, S., Bui, T. Q., Zhang, C., & Lim, C. W. (2014). Numerical modeling of wave propagation in functionally graded materials using time-domain spectral Chebyshev elements. *Journal of Computational Physics, 258*, 381–404.

Olsson III, R. H., El-Kady, I. F., Su, M. F., Tuck, M. R., & Fleming, J. G. (2008) Microfabricated VHF acoustic crystals and waveguides. *Sensors and Actuators A: Physical, 145–146*(0, 7), 87–93.

Olsson III, R. H., & El-Kady, I. F. (2009). Microfabricated phononic crystal devices and applications. *Measurement Science and Technology, 20*(1), 012002.

Chapter 5
Optimisation of Porous 2D PhPs with Respect to In-Plane Stiffness

5.1 Introduction

The optimised topology of 1D bi-material PhPs for maximised RBW of Lamb waves and its variation with respect to filling fraction of constituents was studied in the Chap. 4. This chapter explores the optimised topology of porous PhPs with 2D periodicity for maximised bandgap width of guided waves and incorporates unit-cell's effective stiffness in optimisation algorithm to get structurally worthy porous bandgap topologies.

As discussed in Chap. 1, in-plane symmetric (S) and anti-plane asymmetric (A) Lamb waves as well as symmetric shear horizontal (SHS) and asymmetric shear horizontal (SHA) in-plane waves are the well-known guided wave modes generated in a plate structure. The guided wave dispersion in PhP is governed by its transversal anisotropy i.e. plate's thickness in addition to its planar anisotropy measured by lattice periodicity and unit-cell constitution.

The bandgaps of guided waves in 2D PhPs of prescribed topologies have been extensively studied (Charles et al. 2006; Khelif et al. 2006; Pennec et al. 2010; Wu et al. 2011). PhPs produced by either periodic insertion, attachment or perforation of cylindrical inclusions on a background plate have been generally considered in these studies. Phononic lattice with different patterns e.g. square, rectangular or polygonal have been considered. The phononic unit-cell itself could contain a specific pattern of circular inclusions like the work by Li et al. (2009) who studied guided wave gaps in PhPs with Archimedean-like tiling. However, optimum topology of PhCrs and acoustic meta-materials have been widely investigated, e.g. (Hussein et al. 2006; Krushynska et al. 2014; Manktelow et al. 2013; Olhoff et al. 2012; Rupp et al. 2007).

As discussed in Chap. 2, most of topology optimisation studies in relation to 2D periodic PhCrs have been concerned with bandgaps of in-plane and/or anti-plane bulk waves while only few works have been devoted to bandgaps of guided waves in 2D PhPs. Halkjær et al. (2006) studied the optimum topology of porous

S. Hedayatrasa, *Design Optimisation and Validation of Phononic Crystal Plates for Manipulation of Elastodynamic Guided Waves*, Springer Theses,
https://doi.org/10.1007/978-3-319-72959-6_5

Polycarbonate PhP with rhombic unit-cell for maximised RBW of first couple of flexural (asymmetric) guided waves. The Mindlin's plate theory was implemented for definition of band structure of bending (A + SHA) waves and gradient based optimisation was performed through moving asymptotes method. Since the best topology for maximised RBW did not have acceptable stiffness, new objective was introduced to conversely search widest bandgap width at higher frequency ranges. Finally the discontinuities of optimised topology were locally modified for satisfactory stiffness and manufacturability.

In another investigation by Bilal and Hussein (2012) the optimum topology of thin porous silicon PhPs for maximised RBW of basic flexural waves was studied. The Mindlin's plate theory was implemented and topology of square unit-cell was optimised through GA.

Single material porous unit-cell with uniform through thickness constitution have been commonly considered, in which, the achieved topology could be simply produced by perforation of a uniform background plate. The high acoustic impedance contrast produced by the porosities, usually filled with air, leads to relatively wide acoustic bandgaps. Bandgaps of such porous materials are governed by wave reflection and scattering at the interface of in-homogeneities produced by perforation profile. Therefore the search for highest RBW naturally results in topologies with nearly isolated domains and in other words thin connectivity. The finer the topology resolution the thinner connectivity in the optimised topology for maximised bandgap, unless the topology feature sizes are controlled.

Moreover, porous topology with lower effective stiffness has lower modal frequencies and so lower bandgap frequency range generally leading to higher RBW. Thus largest achievable RBW is extremely dependent on assumed unit-cell's resolution and relevant topology generally has low stiffness. Nevertheless, none of earlier works on topology optimisation of porous PhCrs (Bilal and Hussein 2012; Dong et al. 2014; Halkjær et al. 2006) took into consideration the stiffness of achieved topologies. Although the mesh dependency of topology can be controlled by constraining the minimum length scale of the features, this technique cannot assure the optimality of stiffness. Furthermore, if the topology resolution is not fine enough, this approach may degrade the optimality of bandgap due to imposed feature size constraint to the entire topology area.

In earlier studies the bandgaps of asymmetric guided wave modes have been solely explored (Bilal and Hussein 2012; Dong et al. 2014; Halkjær et al. 2006) while it is also of great value to investigate the topology of PhPs for maximised RBW of symmetric modes as well as complete gap of mixed guided wave modes. Complete bandgap of guided waves can be used as vibration-less supporting structure of delicate dynamic device like gyroscope for immunity to noise sources. Guided waves can be also manipulated through exceptional flat and concave equal frequency contours produced near the bandgap frequency of corresponding mode, for self-collimation and wave steering purposes. If the bandgap is exclusive to either symmetric or asymmetric modes, then the modes can be handled individually. Allowed guided wave modes can be almost purely launched as the banned modes are filtered within the corresponding exclusive bandgaps. Bandgap mode on the other hand can be trapped or guided within appropriate defects in PhP lattice. So, detective wave beams of symmetric or asymmetric modes can be individually excited, steered and captured through PhP elements with promising application in structural health monitoring. Other potential applications are in acoustic energy harvesting and radio frequency (RF) communication (Olsson Iii and El-Kady 2009).

Therefore, in this study the optimum topology of 2D porous PhP with square symmetry is explored for maximised RBW of fundamental (first order) guided wave modes for three different objectives: (i) exclusive bandgap of fundamental asymmetric wave modes A_0 and SHA_0, (ii) exclusive bandgap of fundamental symmetric wave modes S_0 and SHS_0, and (iii) complete bandgap of mixed wave modes. Besides, in-plane stiffness of the porous unit-cell is incorporated as the second objective to be maximised in order to investigate the gradient of bandgap width versus stiffness and to obtain acoustic bandgaps through optimised structural stiffness.

Multi-objective GA NSGA-II is employed to search for the optimum topology taking into account these two objectives. Specialised filtering is applied to the topologies in order to incline the search space towards feasible solutions with no discontinuities without compromising its diversity and randomness. Since the bandgap efficiency of PhP is substantially dependent on its transversal aspect ratio, thick and relatively thin unit-cells are studied separately.

The governing equations, objective functions and developed FEM-GA framework are firstly described in Sect. 5.2. Detailed explanation of FEM modelling of PhP unit-cell as well as adopted GA algorithm and related challenges are discussed in Sect. 5.3. Then achieved optimisation results are presented and discussed in Sect. 5.4. Selected Pareto topologies are given and gradient of RBW and elastic properties are investigated between the two Pareto front extremes (one with widest bandgap and the other one with highest stiffness). Finally, steady state and transient frequency response of finite thin PhP structures of selected Pareto topologies are studied in Sect. 5.5 and validity of obtained results is confirmed.

5.2 Theory and Constitutive Formulations

5.2.1 *Modal Band Analysis of Unit-Cell*

2D PhP is considered to be produced by two dimensional integration of the square basic unit-cell of length a and transversal aspect ratio (width to thickness) a/h, as shown schematically in Fig. 5.1 for an arbitrary perforation profile. The unit-cell is homogeneous along thickness h and heterogeneous in xy-plane due to existence of transversally perforated region(s). The planar anisotropy can lead to constructive scattering of guided waves and consequently bandgaps of guided waves.

Following the procedure already discussed in Chap. 3, the Bloch-Floquet periodic boundary conditions for such plate unit-cell with 2D heterogeneity and traction

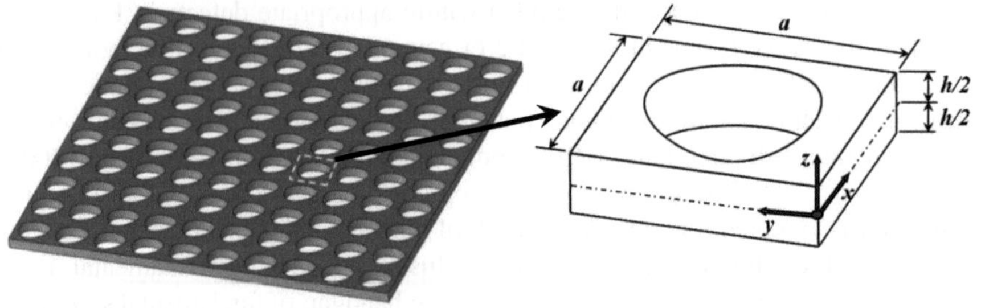

Fig. 5.1 Illustration of a 2D PhP with arbitrary perforation profile, lattice constant a along x and y-axis and thickness h along z-axis

free upper and lower surfaces, can be derived by resolving vector equation Eq. (3.20) into relevant scalar equations:

$$\mathbf{u}_{x+} = \mathbf{u}_{x-}e^{ik_x a} \tag{5.1}$$

$$\mathbf{u}_{y+} = \mathbf{u}_{y-}e^{ik_y a} \tag{5.2}$$

where subscripts x+ and x− stand for corresponding points at two unit-cell's faces parallel to yz-plane at $x = 0$ and $x = a$ respectively. Similarly subscripts y+ and y− stand for corresponding points at two unit-cell's faces parallel to zx-plane at $y = 0$ and $y = a$.

In this study, 3D FEM model of the unit-cell is developed in ANSYS APDL (*ANSYS® Academic Research, Release 14.5*) for calculation of guided waves' modal band structure. Since the Bloch-Floquet boundary conditions (Eqs. 5.1 and 5.2) are complex, two superimposed meshes are modelled so that one represents the real term and the other one imaginary term of complex displacement domain (Aberg and Gudmundson 1997). Consequently the relevant constraint equations connecting two coinciding meshes are expanded as follows for real (Re) and imaginary (Im) terms respectively. So Eq. (5.1) expands to following two equations:

$$(\mathbf{u}_{x+})_{Re} = (\mathbf{u}_{x-})_{Im}\cos(k_x a) - (\mathbf{u}_{x-})_{Re}\sin(k_x a) \tag{5.3}$$

$$(\mathbf{u}_{x+})_{Im} = (\mathbf{u}_{x-})_{Im}\cos(k_x a) + (\mathbf{u}_{x-})_{Re}\sin(k_x a) \tag{5.4}$$

and similarly Eq. (5.2) is expanded to:

$$(\mathbf{u}_{y+})_{Re} = (\mathbf{u}_{y-})_{Im}\cos(k_y a) - (\mathbf{u}_{y-})_{Re}\sin(k_y a) \tag{5.5}$$

$$(\mathbf{u}_{y+})_{Im} = (\mathbf{u}_{y-})_{Im}\cos(k_y a) + (\mathbf{u}_{y-})_{Re}\sin(k_y a) \tag{5.6}$$

So, by applying periodic boundary conditions on vertical (periodic) boundaries and considering natural traction free boundary conditions for upper and lower plate faces, the modal frequencies of the model are calculated versus wave vector over the border of irreducible Brillouin zone to define the band structure of relevant guided waves.

In order to study the band structure of low order Symmetric ($S_0 + SHS_0$) and Asymmetric ($A_0 + SHA_0$) guided waves individually, specific boundary conditions have been considered to stimulate the relevant modes exclusively. This approach relies on the nature of displacement field through the thickness of plate. With regard to harmonic definition of guided waves in uniform plates (Kundu 2004), the harmonic oscillations of a heterogeneous periodic plate in any planar direction in xy-plane, e.g. x-axis, can be explained by symmetric and asymmetric terms with respect to the mid-plane as follows.

Lamb wave oscillations in zx-plane are formulated as:

$$\begin{aligned} u = & \{ikP_1(x)\cos(\alpha_1 z) + \alpha_2 Q_1(x)\cos(\alpha_2 z)\}e^{ikx} \\ & + \{kR_1(x)\sin(\alpha_1 z) + i\alpha_2 S_1(x)\sin(\alpha_2 z)\}e^{ikx} \end{aligned} \tag{5.7}$$

$$\begin{aligned} w = & \{i\alpha P_2(x)\sin(\alpha_1 z) + kQ_2(x)\sin(\alpha_2 z)\}e^{ikx} \\ & + \{\alpha_1 R_2(x)\cos(\alpha_1 z) + ikS_2(x)\cos(\alpha_2 z)\}e^{ikx} \end{aligned} \tag{5.8}$$

based on Stokes-Helmholtz decomposition, and anti-plane oscillations along y-axis are defined by:

$$v = \{P_3(x)\cos(\alpha_3 z)\}e^{ikx} + \{iQ_3(x)\sin(\alpha_3 z)\}e^{ikx} \tag{5.9}$$

where functions P_j, Q_j, R_j and S_j $(j = 1-3)$ are *a* periodic amplitudes of relevant oscillations. Transversal wave numbers α_1, α_2 and α_3 are indeed vertical components of in-plane normal, in-plane shear and anti-plane shear elastic wave numbers respectively, and $k = k_x$.

The first and second terms of Eqs. (5.7), (5.8) and (5.9) apparently stand for symmetric and asymmetric plate modes. So, with regard to traction free surfaces of the plate and explained symmetry of displacement field, these plate modes can be decoupled by applying appropriate boundary conditions to the mid-plane as follows:

$$\{w \quad \tau_{xz} \quad \tau_{yz}\}|_{z=0} = 0 \quad (\mathit{Symetric\ Mode}) \tag{5.10}$$

$$\{u \quad v \quad \sigma_z\}|_{z=0} = 0 \quad (\mathit{Asymetric\ Mode}) \tag{5.11}$$

where σ_z is the transversal normal stress and τ_{xz}, τ_{yz} are transversal shear stresses. Whereas, the displacement field is directly or inversely symmetric with respect to the mid-plane for different polarisations, just half of unit-cell from mid-plane $z = 0$ to top plane $z = h/2$ is modelled for mode decoupling purpose. So, on the plane of symmetry ($z = 0$), the constraint $\{u \quad v\} = 0$ is applied to enforce asymmetric modes, or constraint $w = 0$ is applied to enforce symmetric modes. Zero stress boundary conditions required as per Eqs. (5.10) and (5.11) are naturally satisfied by unconstrained degrees of freedom on the mid-plane in both cases.

5.2.2 Objective Functions

Maximum RBW and highest in-plane stiffness of PhP unit-cell are two opposing objectives of this topology optimisation study. Whereas bandgaps are governed by interfacial reflections of in-homogeneities produced by perforation profile, the search for highest RBW naturally results in topologies with nearly isolated domains and in other words thin connectivity. Finer topology resolution generally results in higher RBW by thinner connectivity, i.e. less stiffness. Therefore, the in-plane stiffness of the PhP unit-cell is introduced as a second objective to study the variation of bandgap efficiency with respect to stiffness. Due to direct contribution of in-plane stresses in through thickness bending moments of a plate, higher in-plane stiffness (of specified thickness) leads to higher overall flexural stiffness as well (Timoshenko and Woinowsky-Krieger 1959). This is particularly valid for the assumed perforated unit-cell with no possibility of through thickness topology and material gradient to control flexural properties. Nonetheless, specific objective function may be defined to achieve particular flexural (bending-torsional) rigidity of interest. However the purpose of current research is to investigate the sensitivity of bandgap efficiency and relevant topology to effective elastic properties in a general sense.

The First (Bandgap) Objective

The first objective function of optimisation is defined as eigenvalue RBW (Eq. 3.28) between lowest possible couple of modal branches j and $j + 1$ to be maximised:

$$F_1 = \Omega_E(j) \tag{5.12}$$

The Second (Stiffness) Objective

As discussed earlier in this section, the second objective of optimisation is to maximise the in-plane stiffness of PhP unit-cell. Principally, the strain energy of the unit-cell under specified boundary tractions represents its relevant stiffness. Higher strain energy implies lower stiffness or higher compliance. The strain energy density W of a finite element subjected to in-plane stress in xy-plane is theoretically defined (Kaw 2005) as:

$$W = \frac{1}{2}(\sigma_x \varepsilon_x + \sigma_y \varepsilon_y + \tau_{xy}\gamma_{xy}) \tag{5.13}$$

where $\{\sigma_x \quad \sigma_y \quad \tau_{xy}\}$ and $\{\varepsilon_x \quad \varepsilon_y \quad \gamma_{xy}\}$ are corresponding in-plane stress and strain vectors respectively. Now by assumption of a general in-plane stress state:

$$\{\, \sigma_x \quad \sigma_y \quad \tau_{xy} \,\} = \{\, \sigma \quad \sigma \quad b_s \quad \sigma \,\} \tag{5.14}$$

and with regard to linear elastic stress-strain relations, Eq. (5.13) can be rewritten as:

$$W = \sigma^2 (\frac{1 - \nu_e}{E_e} + \frac{b_s^2}{2G_e}) \tag{5.15}$$

where E_e, G_e and ν_e are the effective in-plane orthotropic elastic modulus, shear modulus and Poisson's ratio of assumed square symmetric PhP unit-cell respectively, and b_s in Eq. (5.14) is *stress ratio* defining the relative magnitude of shear to normal stresses. Finally, the second objective function F_2 is defined as the relative strain energy compliance of unit-cell to be minimised:

$$F_2 = \frac{1}{\epsilon_s}\left(\left(\frac{1 - \nu_e}{E_e}\right) + \left(\frac{b_s^2}{2G_e}\right)\right) \tag{5.16}$$

where constant ϵ_s is the compliance of pure solid background:

$$\epsilon_s = \left(\frac{1 - \nu_s}{E_s}\right) + \left(\frac{b_s^2}{2G_e}\right) \tag{5.17}$$

The ratio of longitudinal compliance to shear compliance and their priority could be controlled by the stress ratio b_s. Various techniques have been developed in the literature to define the equivalent properties of heterogeneous materials. The methods basically define the equivalent properties of a representative volume element which herein is the unit-cell of phononic lattice. Asymptotic homogenisation is a rigorous approach in defining the effective properties of composite materials and provides seamless estimations of microstructural fluctuations of stress and strain through decoupling macro and micro deformations (Oliveira et al. 2009; Pinho-da-Cruz et al. 2009). Specialised high order multiscale homogenisation approaches have also been developed to define the equivalent material properties and wave dispersion of periodic lattice structures (Gonella and Ruzzene 2008; Oliveira et al. 2009).

However the overall macromechanical stiffness of the PhP unit-cell is required to be maximised herein. For this purpose, the standard numerical homogenisation approach is employed to define the equivalent elastic properties of PhP unit-cell and calculate F_2, which has been reliably used for FEM based homogenisation of composite materials (Berger et al. 2005; Chen et al. 2016; Kari et al. 2007a, b;

Steven 2006; Xia et al. 2003). This approach is based on mechanics of materials and defines the homogenised properties by equalising the strain energy of heterogeneous unit-cell and that of uniform equivalent material, under prescribed test strains or stresses.

It has been shown that neither uniform displacements nor uniform traction forces on the boundaries can appropriately prescribe the real boundary conditions of unit-cell as a subset of a periodic lattice structure. The first approach overestimates and the latter approach underestimates the homogenised properties. Periodic boundary conditions which correlate corresponding points of opposite boundaries (rather than a uniform correlation of opposite faces), define and maintain the periodicity of unit-cell the best (i.e. no overlap of neighbouring unit-cells after deformation) and provide the most accurate homogenised properties (Hollister and Kikuchi 1992; Xia et al. 2003).

Therefore, periodic boundary conditions are applied and overall test strain $\bar{\boldsymbol{\varepsilon}}$ is applied to the unit-cell boundaries. Then the equivalent orthotropic stiffness $\mathbf{D}_e$ is derived based on volumetric average stress $\bar{\boldsymbol{\sigma}}$ of elements over the entire unit-cell:

$$\bar{\boldsymbol{\sigma}} = \frac{1}{V}\int_V \boldsymbol{\sigma} dV \tag{5.18}$$

$$\mathbf{D_e} = \bar{\boldsymbol{\sigma}}\,\bar{\boldsymbol{\varepsilon}}^{-1} = \begin{bmatrix} C_{11} & C_{12} & C_{13} & 0 & 0 & 0 \\ C_{12} & C_{11} & C_{13} & 0 & 0 & 0 \\ C_{13} & C_{13} & C_{33} & 0 & 0 & 0 \\ 0 & 0 & 0 & C_{44} & 0 & 0 \\ 0 & 0 & 0 & 0 & C_{55} & 0 \\ 0 & 0 & 0 & 0 & 0 & C_{55} \end{bmatrix} \tag{5.19}$$

Four different strain fields with pure longitudinal strain $\bar{\varepsilon}_x$, pure transversal strain $\bar{\varepsilon}_z$, and pure shear strains $\bar{\tau}_{xy}$ and $\bar{\tau}_{xz}$ are required to define independent homogenised stiffness constants of this square symmetric topology. However, in-plane test strains $\bar{\varepsilon}_x$ and $\bar{\tau}_{xy}$ suffice for calculation of the required in-plane engineering elastic properties E_e, G_e and ν_e (Eq. 5.16) for plate unit-cell with traction free lateral faces. The selection of test strains' magnitude is arbitrary and small strains of 0.001 are applied herein.

The dispersion-less (low frequency) propagation speed of S_0 and SH_0 plate modes determined from the initial slope of relevant dispersion curves of unit-cell are governed by in-plane longitudinal and shear stiffness of the unit-cell (Kundu 2004). One may alternatively consider them as representatives of unit-cell's effective stiffness in definition of objective function with less computational effort. However, it has complications in orthotropic medium and in this work the numerical homogenisation is employed as a more generic approach.

5.3 Modelling Parameters for FEM Analysis

Polysilicon with material properties $E_s = 169\,\mathrm{GPa}$, $\nu_s = 0.22$ and $\rho_s = 2330\,\mathrm{kg/m^3}$ (Sharpe et al. 1997) is taken as the solid background material of PhP in this work which has been widely used as structural material for fabrication of micro-devices like micro PhCrs (Olsson Iii and El-Kady 2009). A unit-cell with element resolution 32 × 32 in xy-plane is selected leading to a design domain with 136 independent design variables for square symmetric topology. Two transversal aspect ratios a/h = 2 and 10 are considered in order to investigate the design and efficiency of thick and relatively thin plate unit-cells respectively. ANSYS linear solid element SOLID185 is used for FEM modelling of thick unit-cells, and solid-shell element SOLSH190 is used for modelling thin unit-cells. The transversal mesh resolution and challenges in handling void sections of porous topology are detailed in the following sections.

5.3.1 *Transversal Mesh Resolution*

The selected planar mesh resolution 32 × 32 showed to provide adequate accuracy for FEM analysis of fundamental modal frequencies of interest, given that appropriate number of element layers is considered through the thickness of unit-cell. Since GA optimisation demands FEM analysis to be performed for a large number of iterations, it is important to develop a reliable model with minimised computational intensity.

Proper mesh resolution along the thickness of unit-cell h relies on its transversal aspect ratio and the frequency range of highest guided wave mode (to be calculated). SOLSH190 element usually gives more accurate predictions compared to classical shell elements (with Mindlin's Plate formulation) when the shell is thick (*ANSYS 14.5 element reference*). Thus, for modelling moderately thick PhP unit-cells of aspect ratio 10 a single layer of solid-shell elements provides good accuracy for modal band analysis of low order modal branches. As regard to thick unit-cell with aspect ratio 2, Fig. 5.2 displays the modal frequencies at e.g. Brillouin zone point X versus the number of element layers along the thickness obtained for an arbitrary topology shown in Fig. 5.2a and material property ratio $P_s/P_v = 1000$ discussed later in Sect. 5.3.2. Figure 5.2a, b show the first 10 mixed guided wave modes and first 10 asymmetric modes respectively (9 modes visible due to coincidence).

Obviously, the steepest ramp occurs when increasing element layers from one to two, and the calculated asymmetric modes are less sensitive to transversal mesh resolution (Fig. 5.2b). This is due to the fact that just one half of plate's thickness is modelled for mode decoupling which provides more accuracy compared to the model with one element layer for the entire thickness. In both cases the highest mode converges well after considering up to 10 layers. Although increasing the

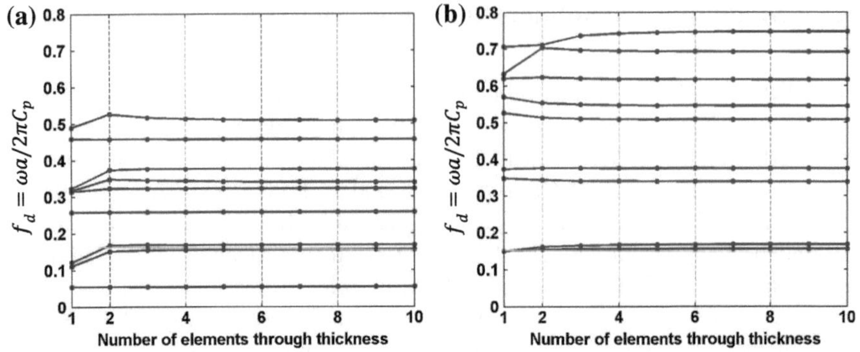

Fig. 5.2 Gradient of first few modal frequencies versus number of element layers through thickness of connected topology shown in Fig. 5.3a for transversal aspect ratio $a/h = 2$ and $P_s/P_v = 1000$; **a** whole unit-cell modelled for mixed plate modes analysis, and **b** half of unit-cell's thickness modelled from the mid-plane for decoupling asymmetric modes

number of element layers leads to more accurate modal analysis, considering 1 element layer proved to be adequate for estimation and widening of lowest bandgap during optimisation.

Two element layers are considered only for the unit-cell of aspect ratio 2 when maximising RBW of complete guided wave's gap. For calculation of the modal band structure and RBW of achieved topologies higher mesh resolutions of 10 and 4 layers are considered for aspect ratios 2 and 10 respectively.

5.3.2 Modelling Void Segments and Compliant Material Properties

For FEM evaluation of topologies, only the solid part of topology has to be ideally modelled and appropriate boundary conditions should be applied to existing boundaries at any evaluation iteration. This approach is challenging to be included in an automated algorithm like GA and of course its complexity depends on the capabilities of implemented FEM solver. Also, as a general simplified approach a compliant material can be assigned to the void elements by a few orders less than the solid background. However, the degradation order of solid background properties to be considered as void and the constitution of randomly generated topologies are crucial factors in reliable calculation of RBW.

In order to investigate this matter, the sensitivity of modal frequencies of plate unit-cell to the material property ratio P_s/P_v is studied as shown in Fig. 5.3. The variables P_s and P_v stand for material properties (density or elastic modulus) of solid and compliant void phases respectively. Unit-cell with aspect ratio of 2 and transversal mesh resolution of 10 is modelled for this purpose. Both elastic modulus

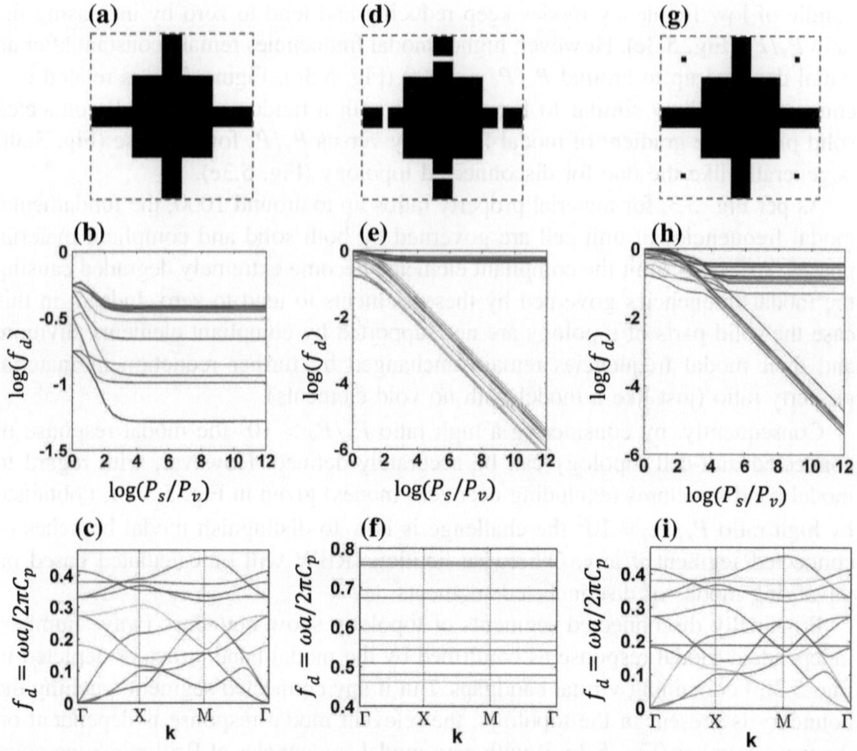

Fig. 5.3 Topology and related modal response of unit-cell at X(i.e. $\mathbf{k} = [\pi/a \quad 0]$) versus ratio of material properties P_s/P_v, as well as the modal band structure of unit-cell for $P_s/P_v = 10^9$; **a**, **b**, **c** connected topology; **d**, **e**, **f** disconnected topology; and **g**, **h**, **i** connected topology including disconnected segments

and density of solid material are reduced with the same order to be considered as compliant material. So extraordinary mass to stiffness ratio leading to virtual low modal frequencies is avoided (Bendsoe and Sigmund 2003).

With regard to the random nature of GA optimisation and relevant individual topologies to be analysed during fitness evaluation process, three distinct arbitrary topologies are studied as depicted in Fig. 5.3a, d, g). Then the gradient of dimensionless modal frequency ω_d versus P_s/P_v at arbitrary Brillouin zone point X (i.e. $\mathbf{k} = [\pi/a \quad 0]$) is presented in logarithmic scale in Fig. 5.3b, e, h for these topologies respectively. The first topology illustrated in Fig. 5.3a contains just one solid inclusion of filling fraction $v_f = 0.375$ by which obviously a connected lattice plate could be produced. According to the relevant results in Fig. 5.3b the modal frequencies decrease dramatically until around $P_s/P_v = 1000$ and after slight variation up to around $P_s/P_v = 10^5$ show absolutely no variation any further. But, by insertion of an arbitrary disconnection in the topology as shown in Fig. 5.3d, a bundle of low frequency modes keep reducing and tend to zero by increasing the ratio P_s/P_v (Fig. 5.3e). However, higher modal frequencies remain constant after an initial descend up to around $P_s/P_v = 1000$ (Fig. 5.3e). Figure 5.3g is related to a connected topology similar to Fig. 5.3a but with a randomly added disconnected solid pixel. The gradient of modal frequency versus P_s/P_v for this case (Fig. 5.3h) is generally like the one for disconnected topology (Fig. 5.3e).

As per Fig. 5.3, for material property ratios up to around 1000, the fundamental modal frequencies of unit-cell are governed by both solid and compliant material phases. After this limit the compliant elements become extremely degraded causing the modal frequencies governed by these elements to tend to zero. Indeed, in this case the solid parts of topology are not supported by compliant elements anymore and their modal frequencies remain unchanged by further reduction of material property ratio (just like a model with no void elements).

Consequently, by considering a high ratio $P_s/P_v > 10^5$ the modal response of connected unit-cell topology can be accurately defined. However, with regard to modal band structures (excluding near zero modes) given in Fig. 5.3c, f, i obtained by high ratio $P_s/P_v = 10^9$ the challenge is how to distinguish modal branches of connected segment if any. Otherwise fictitious RBW will be calculated based on unvarying modes of disconnected segments.

Principally disconnected segments of topology show boundary (wave number) independent modal response as confirmed by the modal band structure depicted in Fig. 5.3f, i containing virtual bandgaps. But if any connected segment touching the boundary is present in the topology, the relevant modal response is dependent on the wave number (Fig. 5.3c, i) with zero modal frequencies at Brillouin zone point Γ (i.e. $\mathbf{k} = [0 \quad 0]$) concerning the rigid body motion in three degrees of freedom.

Considering relatively low material property ratio of around 1000 generally provides a good approximation of modal band structure for connected unit-cell in the absence of instability and near zero modal frequencies. But, using this low ratio leads to impractical optimised topologies comprised of isolated segments predominantly affecting the produced bandgap (Halkjær et al. 2006). Actually the problem converges towards topologies producing bandgaps by locally resonant solid inclusions held by surrounding weak compliant elements.

Hence, for accurate modal band analysis of the porous unit-cell, sufficiently high degradation ratio $P_s/P_v > 10^5$ has to be considered, or void elements and relevant boundary conditions have to be removed from the model. In any case, a robust topology recognition and processing as implemented in present study is essential to assure the feasibility and validity of optimised topologies.

In this part of research the density and elastic modulus of solid background are reduced by 9 orders of magnitude to be considered as void material. However, in further steps of research presented in the following chapters, the ANSYS APDL script was improved to efficiently remove void elements and relevant boundary conditions before the solution stage. This improvement had considerable impact on computational efficiency of fitness evaluation and overall optimisation time.

5.4 NSGA-II Optimisation Algorithm

The following adopted NSGA-II algorithm is coded in MATLAB (*MATLAB 8.0, The MathWorks Inc.*) which iteratively executes the developed ANSYS macro file for FEM modelling and fitness evaluation purpose. Topology with square symmetry is prescribed for assumed square shape of PhP unit-cell and a binary bit-string design space is considered for defining the material type of any pixel as discussed in Chap. 3. Specialised topology initialising, filtering and modifications are applied to eliminate interrupting modes and make optimised topologies converge towards feasible connected ones without disturbing the randomness and diversity of search domain. The adopted NSGA-II algorithm covering these techniques is as follows:

1. Create a random bit-string initial population for the solid-void problem defining the material type at each column of elements in FEM model as a design variable (gene). An identical fully solid population is randomly mutated with specified probability 0.2 to deliver an initial population of likely connected topologies.
2. Accept individuals for fitness evaluation provided that the relevant topology *includes* valid segment touching boundary and with face to face connected elements. Else, degrade the topology by assigning very low fitness for both objectives.
3. Determine the objective functions for accepted individuals by considering the valid boundary connected segment *only*. The segments having single node (hinge) connection to the boundary connected segment are retained for further evolution.
4. If the bandgap objective $F_1 \leq 0$ (i.e. no bandgap) then totally compliant material (void) is considered to define stiffness objective F_2. In this way the Pareto front delivers a spread of bandgap topologies merely.
5. Sort the population based on the rank and crowding distance of individuals (Pratap et al. 2002).
6. Create parent pool through tournament selection. Two randomly selected individuals are compared and the one with higher rank is selected as candidate parent. If both have the same rank, the one with highest crowding distance is selected.
7. Create offspring population through single point crossover with certain probability. Two children are produced per couples of randomly selected parents.
8. Mutate gens of offspring population with specified probability.
9. Modify topology for checkerboard elimination. Randomly selected pixel is inversed (i.e. 0 to 1 and vice versa) if surrounded by more than 2 dissimilar adjacent face to face pixels out of 4. This well-known technique alleviates the checker board problem which leads to overestimated stiffness when linear elements are used.
10. Modify topology in order to eliminate invalid disconnected segments of topology. Randomly selected disconnected segment (including any segment with hinge connection) is eliminated with specified probability.

11. Produce intermediate population by accumulating offspring and current population and sort them based on ranking and crowding distance.
12. Select individuals for the next generation first based on their ranking and second, if the first rank is bigger than required population based on their crowding distance.
13. Go to Step-2 if termination condition (prescribed number of generations) is not satisfied, otherwise stop.

Due to rarity of perfectly connected topologies showing bandgap properties defined during population initialisation and reproduction stages, the aforementioned algorithm is a robust approach for filtering invalid individuals while maintaining the diversity of search domain. Among other alternatives (i) fitness penalisation disturbs optimality of Pareto front, (ii) accepting perfectly connected topologies restricts the diversity of search space, and (iii) filtering individuals to acceptable ones interferes with the randomness of GA search. In fact, in the introduced approach existence of any disconnected segment in the topology does not interrupt its fitness evaluation and let it contribute to the evolution process. Single node hinge connection is also avoided. Actually this weak form of connection provides segments with degraded or virtually stiff bonding which opens wide bandgaps and therefore relevant topologies usually stand as elite solutions in the first rank. MATLAB Image Processing Toolbox (*MATLAB 8.0, The MathWorks Inc.*) is employed to perform required topology filtering and morphology assessments.

The mutation of offspring as mentioned in Step-7 of the algorithm manipulates the next generation with specified probability. This probability proved to remarkably affect the convergence rate and efficiency of GA search by retaining diversity and preventing premature convergence to a local optimum. This is due to infrequency of usable topologies opening bandgap particularly between lowest modes of interest. Moreover, the effective mutation probability depends on the filling fraction of high ranked topologies. Lower filling fraction means more void segments with less effectiveness probability on the fitness when inversed via mutation and so requiring higher mutation to exploit new connected designs. High mutation probability on the other hand should be avoided as it disturbs the genome and prevents convergence to a unique optimised topology.

Population size 200 is chosen for optimisation studies and GA operation probabilities are varied during optimisation in 4 steps as given in Table 5.1. In Stage-1 the design space is coarsely searched by high mutation probability 0.3 for only 5 generations to have a good resource of initiating population. Then lower mutation probability 0.1 is used for another 40 generations which greatly mitigates the risk of getting stuck in a local optimum.

Afterwards, low mutation probability 0.05 is kept for up to 75 new generations. The segment elimination probability is added and increased gradually to avoid disturbance of genomes in the early stages of optimisation. In fact, any disconnected segment potentially can fill a considerable fraction of unit-cell and so initial intense elimination can make the GA search turn biased. Finally, the optimisation is continued for a number of generations with 100% segment elimination in the

Table 5.1 Employed GA settings for topology optimisation studies

Stage	Probability			Generations
	Crossover	Mutation	Disconnected segment elimination	
1	0.9	0.3	0.0	5
2	0.9	0.1	0.2	40
3	0.9	0.05	0.5	75
4	0.9	0.0	1.0	20–50

absence of mutation to get refined feasible topologies with checkerboard improvements.

5.5 Topology Optimisation Results

The topology optimisation results obtained through the aforementioned multi-objective GA are delivered in this section. The optimum design of PhP unit-cell is achieved for 3 different bandgap objectives, as follows:

1. Bandgap Objective-1: to maximise RBW of lowest gap of fundamental asymmetric guided wave modes including asymmetric lamb mode A_0 and asymmetric shear horizontal mode SHA_0
2. Bandgap Objective-2: to maximise RBW of lowest gap of fundamental symmetric guided wave modes including symmetric Lamb mode S_0 and symmetric shear horizontal mode SHS_0
3. Bandgap Objective-3: to maximise RBW of lowest complete gap of mixed fundamental guided wave modes (S_0, A_0, SHS_0 and SHA_0)

These are separately considered as the first objective of optimisation, as well as considering maximised in-plane stiffness as the second objective. Two transversal aspect ratios 2 and 10 are considered to study the design and efficiency of thick and relatively thin plate unit-cells respectively.

In any optimisation case, a set of 10–12 solutions are selected from Pareto front containing a variety of alternative optimised topology modes with different RBWs and in-plane stiffness. The two extreme topologies of Pareto front with widest bandgap and highest stiffness are included; however stiffest topology with very minor bandgap is ignored. The selected optimised Pareto topologies for 8 different objective settings are referred to as TA to TH, to be discussed later on.

Whereas frequency RBW (Ω_F) and eigenvalue RBW (Ω_E) have been used in presenting the bandgap efficiency of PhCrs in different works (Bilal and Hussein 2012; Halkjær et al. 2006), both quantities are obtained and included in the results. In order to evaluate the relative effective stiffness of optimised topologies with respect to stiffness of constitutive solid material, relative elastic modulus $E_r = E_e/E_s$ and relative shear modulus $G_r = G_e/G_s$ are calculated. For non-dimensional modal analysis of unit-cell, dimensionless angular frequency is defined as $\omega_d = \omega a / C_s$ where $C_s = \sqrt{E_s/\rho_s} = 8516\,\text{m/s}$ based on material properties of pure Polysilicon.

5.5.1 Thin PhP with Aspect Ratio 10: Bandgap Objective-1 and Effect of Stress Ratio b_s

As for the PhP of aspect ratio 10, firstly the results concerning Bandgap Objective-1 are presented to maximise RBW of lowest possible gap between the first couple of asymmetric modal branches. Three different stress ratios $b_s = \sqrt{2}$, 0 and 2 are examined in this case. With regard to Eq. (5.16) in relation to the stiffness objective of optimisation F_2, the stress ratio $b_s = \sqrt{2}$ imposes even contribution of normal and shear stiffness. Other values 0 and 2 obviously lead to pure contribution of normal stiffness and weighted dominant contribution of shear stiffness respectively. Although the dominancy of normal or shear stiffness of final solution depends on the nature of bandgap topologies, the Pareto front is inclined to either side through defined value of b_s.

The Pareto front of optimisation for Bandgap Objective-1 and $b_s = \sqrt{2}$, and relative location of selected optimised topologies TA1 to TA12 are shown in Fig. 5.4. These selected Pareto front topologies are shown in Fig. 5.5a. The gradient of RBW, filling fraction v_f and homogenised Poisson's ratio v_e across these Pareto front topologies are given in Fig. 5.5b. The variation of relative elastic

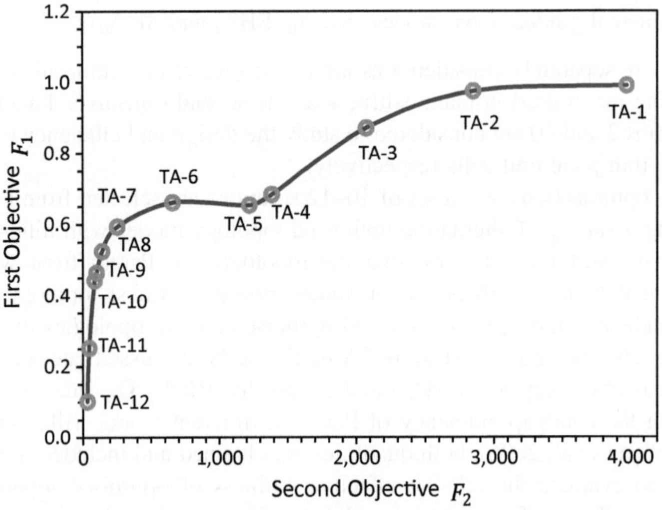

Fig. 5.4 Pareto front of optimisation for bandgap objective-1 and stress ratio $b_s = \sqrt{2}$, and selected optimised topologies TA1 to TA12 (Fig. 5.5a) across its two extremes

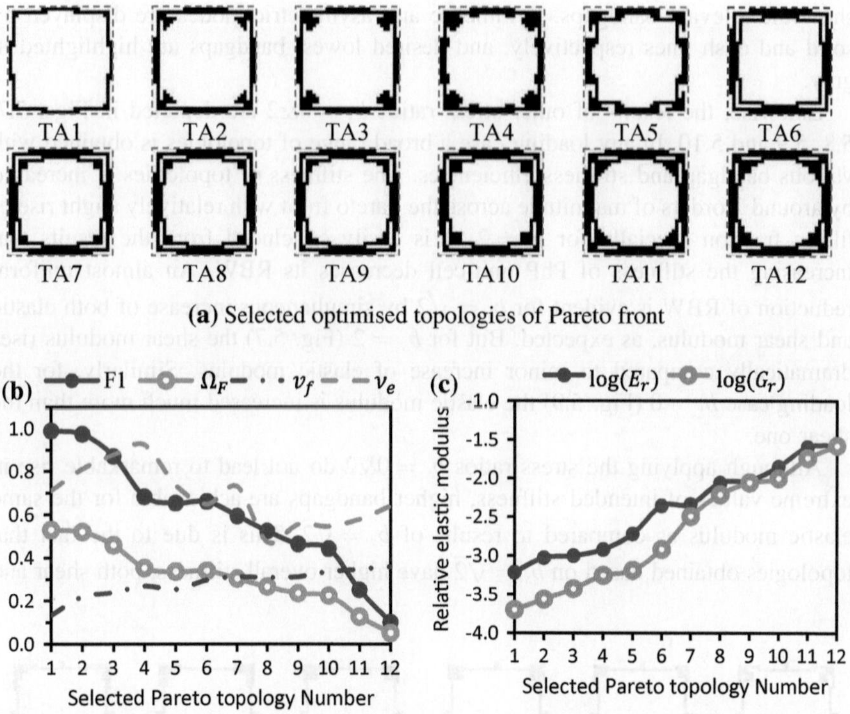

Fig. 5.5 Topology optimisation results of PhP unit-cell with aspect ratio 10 for maximised bandgap objective-1, stress ratio $b_s = \sqrt{2}$; optimised topologies TA1 to TA12 are selected from relevant Pareto front (Fig. 5.4)

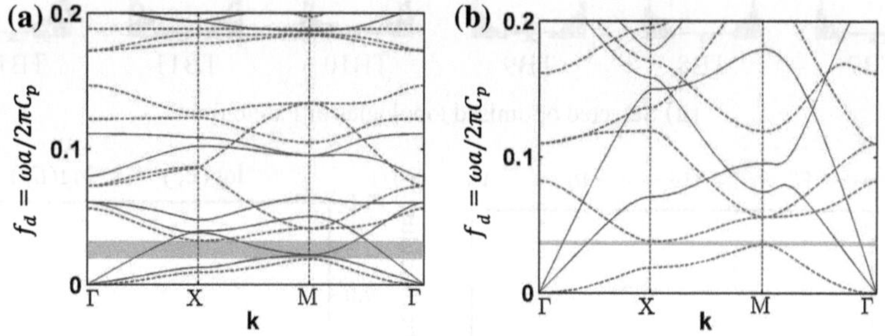

Fig. 5.6 Modal band structure of guided wave modes for **a** topology TA1, and **b** topology TA12 (Fig. 5.5); symmetric and asymmetric modes are shown by solid and dash lines respectively, and lowest bandgap of asymmetric modes is highlighted in grey

properties are also given in Fig. 5.5c in logarithmic scale to define the magnitude order of relative elastic modulus of selected topologies. Moreover, the modal band structure of two extreme topologies of Pareto front with maximum RBW (TA1) and maximum in-plane stiffness (TA12) are depicted in Fig. 5.6a, b respectively, to show the relevant bandgaps. Symmetric and asymmetric modes are displayed by solid and dash lines respectively, and desired lowest bandgaps are highlighted in grey.

Likewise, the results of other stress ratios $b_s = 0\&2$ are depicted in Figs. 5.7, 5.8, 5.9 and 5.10. In any loading case a broad range of topologies is obtained with various bandgap and stiffness efficiencies. The stiffness of topologies is increased by around 2 orders of magnitude across the Pareto front with relatively slight rise of filling fraction specially for $b_s = 2$. It is easily concluded from the results that increasing the stiffness of PhP unit-cell decreases its RBW. An almost uniform reduction of RBW is evident for $b_s = \sqrt{2}$ by simultaneous increase of both elastic and shear modulus, as expected. But for $b_s = 2$ (Fig. 5.7) the shear modulus rises dramatically compared to minor increase of elastic modulus. Similarly, for the loading case $b_s = 0$ (Fig. 5.9) the elastic modulus is increased much more than the shear one.

Although applying the stress ratios $b_s = 0\&2$ do not lead to remarkable rise in extreme values of intended stiffness, higher bandgaps are achievable for the same elastic modulus as compared to results of $b_s = \sqrt{2}$. This is due to the fact that topologies obtained based on $b_s = \sqrt{2}$ have higher overall stiffness, both shear and

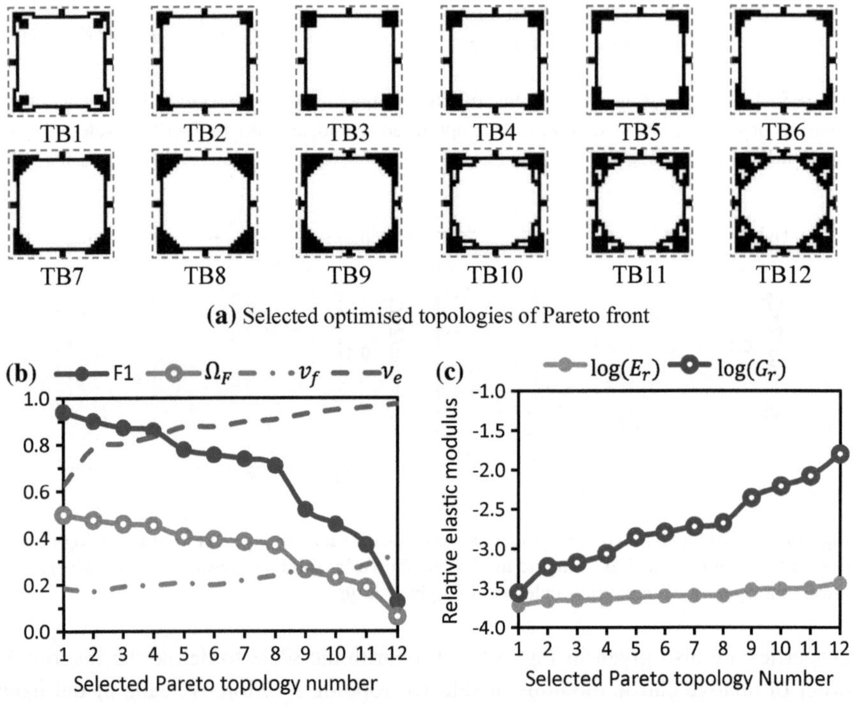

Fig. 5.7 Topology optimisation results of PhP unit-cell with aspect ratio 10 for maximised bandgap objective-1, stress ratio $b_s = 2$; optimised topologies TB1 to TB12 are selected from relevant Pareto front

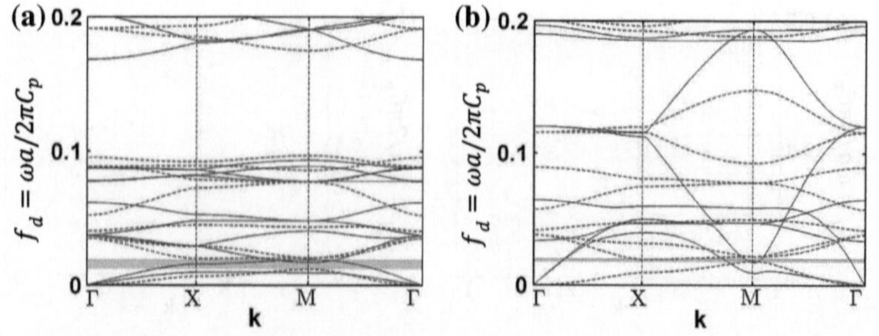

Fig. 5.8 Modal band structure of guided wave modes for **a** topology TB1, and **b** topology TB12 (Fig. 5.7); symmetric and asymmetric modes are shown by solid and dash lines respectively, and lowest bandgap of asymmetric modes is highlighted in grey

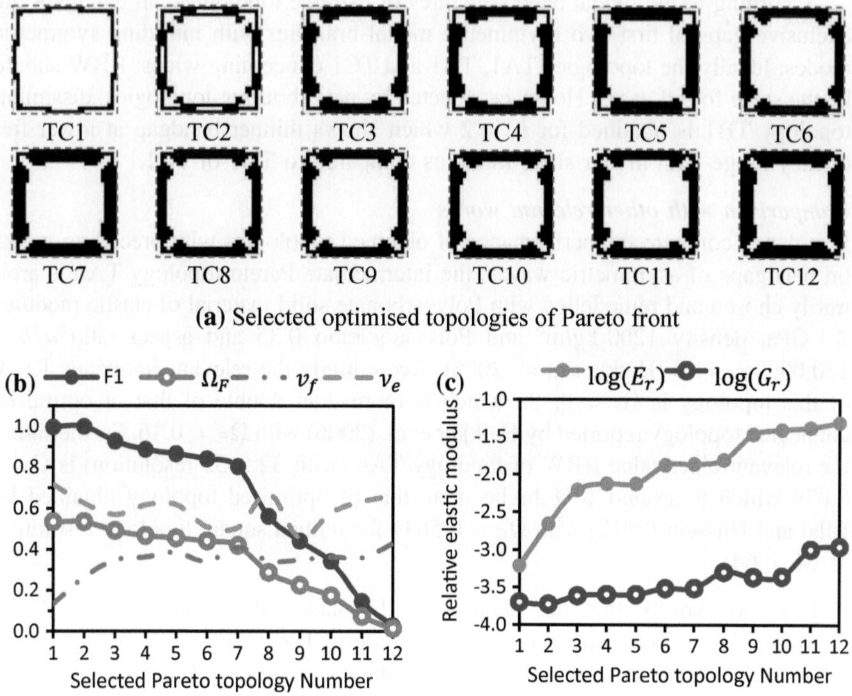

Fig. 5.9 Topology optimisation results of PhP unit-cell with aspect ratio 10 for maximised bandgap objective-1, stress ratio $b_s = 0$; optimised topologies TC1 to TC12 are selected from relevant Pareto front

normal, which degrades the bandgap efficiency. However, the optimality and supremacy of topologies depend on the desired functionality of PhP. Alternatively the stress ratio $b_s = \sqrt{2}$ is considered for all remaining optimisation cases of present work to get almost equally normal-shear stiffened topologies.

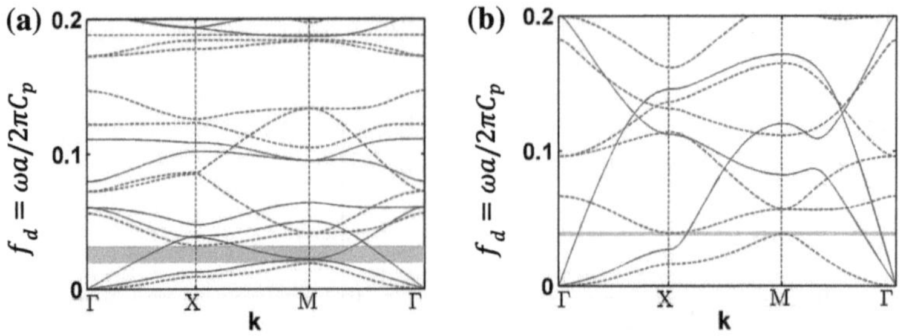

Fig. 5.10 Modal band structure of guided wave modes for **a** topology TC1, and **b** topology TC12 (Fig. 5.9); symmetric and asymmetric modes are shown by solid and dash lines respectively, and lowest bandgap of asymmetric modes is highlighted in grey

According to the modal band structures of extreme topologies, all bandgaps are exclusive gaps of first two asymmetric modal branches with intruding symmetric modes. Ideally the topologies TA1, TB1 and TC1 concerning widest RBW should be the same for all cases. However, affected by neighbouring topologies, dissimilar topology TB1 is obtained for $b_s = 2$ which shows thinner bandgap at lower frequency range with higher shear modulus compared to TA1 or TC1.

Comparison with other relevant works

In order to compare the performance of obtained topologies with preceding works on bandgaps of asymmetric waves, the intermediate Pareto topology TA6 is arbitrarily chosen and remodelled with Polycarbonate solid material of elastic modulus 2.3 GPa, density 1200 kg/m^3 and Poisson's ratio 0.35 and aspect ratio $a/h = 1/0.0909 = 11.0$ (Halkjær et al. 2006). Accordingly the relevant frequency RBW of this topology is $\Omega_F = 0.328$ which is more than double of that of optimised connected topology reported by Halkjær et al. (2006) with $\Omega_F = 0.16$. Furthermore, the relevant eigenvalue RBW of topology TA6 (with 32 × 32 resolution) is $\Omega_E = 0.639$ which is around 13% higher than that of optimised topology obtained by Bilal and Hussein (2012) with $\Omega_E = 0.5639$ for refined square topology resolution of 64 × 64.

It is noteworthy that the bandgap performance of an intermediate Pareto topology (TA6), and even not the one with widest RBW (TA1), was compared with other works aimed at topology with absolutely maximised RBW. Moreover, the topologies of present study are optimised based on slightly different aspect ratio $10 < 11$. However this comparison can only show the competency of implemented approach in digging out the desired optimum bandgap topology. Whereas the RBW of PhP is highly dependent on its in-plane stiffness, a fair judgment about relative bandgap-structural performance of referred topologies demands their in-plane stiffness to be evaluated and taken into account.

5.5.2 Thin PhP with Aspect Ratio 10: Bandgap Objectives 2 and 3

Alternatively, the stress ratio $b_s = \sqrt{2}$ is considered for all remaining optimisation cases of present work to get almost equally normal-shear stiffened topologies. Regarding Bandgap Objective-2 the search for lowest feasible bandgap of symmetric modes between 2nd and 3rd modal branches led to barely opened gap through limited spread of topologies. The best Pareto topology with maximum RBW of $\Omega_E = 0.166$ and its modal band structure is given in Fig. 5.11, which show how the desired gap is opened by means of a locally resonant inclusion with weak (hinge) connection to the surrounding solid frame.

Due to discreetness of the wave vectors on the border of Brillouin zone to be evaluated (30 points in this study) it is impossible to capture the sharp approaching points of the two modal branches located away from the search points. This deficiency leads to invalid or overestimated bandgaps when the modes are tangential or gap width is tiny, like all Pareto topologies with face to face connection achieved for this case. Strengthened connectivity can also be obtained by modifying the bandgap objective to search for gaps at higher midgap frequencies [e.g. the work by Halkjær et al. (2006)], though the desired low frequency performance of produced gap is compromised in this way.

Hence, preferably, the succeeding gap between 3rd and 4th symmetric modal branches is investigated to be maximised as the Bandgap Objective-2.

The optimisation results of this case for stress ratio $b_s = \sqrt{2}$ are given in Fig. 5.12. Obviously, significantly higher RBW is obtained through wide spread of well-connected topologies. The lowest frequency of bandgap is also less than what obtained for 2nd and 3rd modal branches. The elastic modulus is increased by

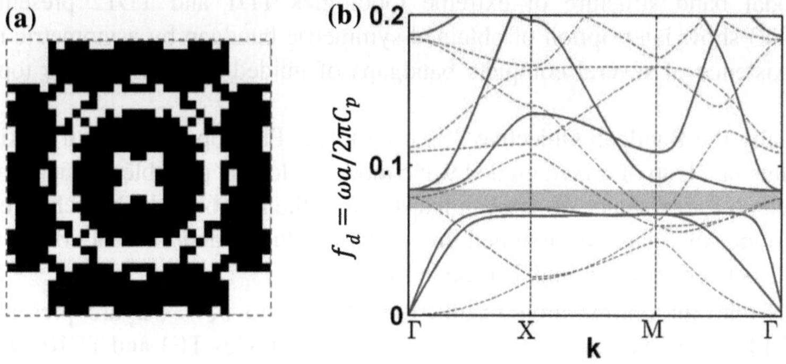

Fig. 5.11 **a** Best Pareto topology for maximised RBW between 2nd and 3rd modal branches of symmetric guided wave modes and **b** relevant modal band structure; symmetric and asymmetric modes are shown by solid and dash lines respectively, and bandgap of symmetric modes is highlighted in grey

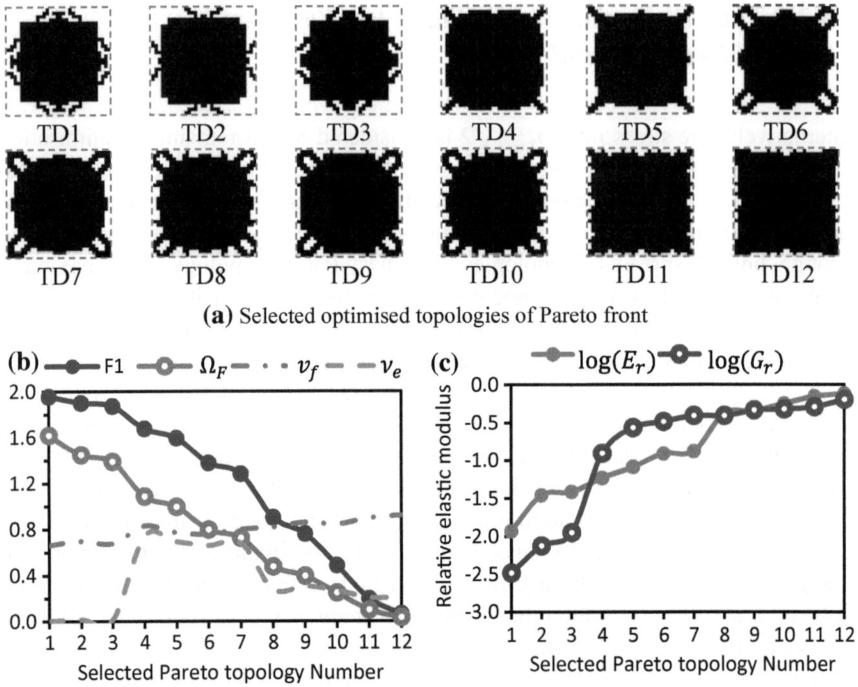

Fig. 5.12 Topology optimisation results of PhP unit-cell with aspect ratio 10 for maximised bandgap objective-2, stress ratio $b_s = \sqrt{2}$; optimised topologies TD1 to TD12 are selected from relevant Pareto front

almost two orders of magnitude across the Pareto front within the filling fraction range of 0.66–0.92.

Modal band structure of extreme topologies TD1 and TD12 presented in Fig. 5.13 show interruption of obtained symmetric bandgap by asymmetric modes and existence of several complete bandgaps of guided wave modes for topology TD1.

Finally the Bandgap Objective-3 is used to get PhP topologies with maximised complete bandgap of mixed guided wave modes at lowest possible frequency range. The search for existence of any bandgaps within the first 6 modal branches resulted in no noticeable gaps for assumed unit-cell's resolution and the first complete gap opened between the 6th and 7th modal branches.

The relevant optimisation results for stress ratio $b_s = \sqrt{2}$ are presented in Fig. 5.14 and modal band structure of extreme topologies TE1 and TE10 are given in Fig. 5.15. Compared to Bandgap Objectives-1 and 2, considerably lower spread of topologies is achieved in the Pareto front around the filling fraction 0.5. This could be due to rarity of topologies with bandgaps of both symmetric and asymmetric modes. Both normal and shear elastic modulus are increased by around 0.5 order of magnitude over the entire range, regardless of the stiffest topology TE10

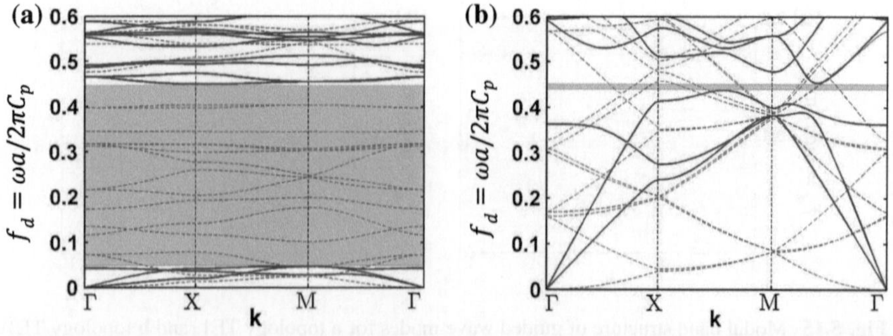

Fig. 5.13 Modal band structure of guided wave modes for **a** topology TD1, and **b** topology TD12 (Fig. 5.12); symmetric and asymmetric modes are shown by solid and dash lines respectively, and lowest bandgap of symmetric modes is highlighted in grey

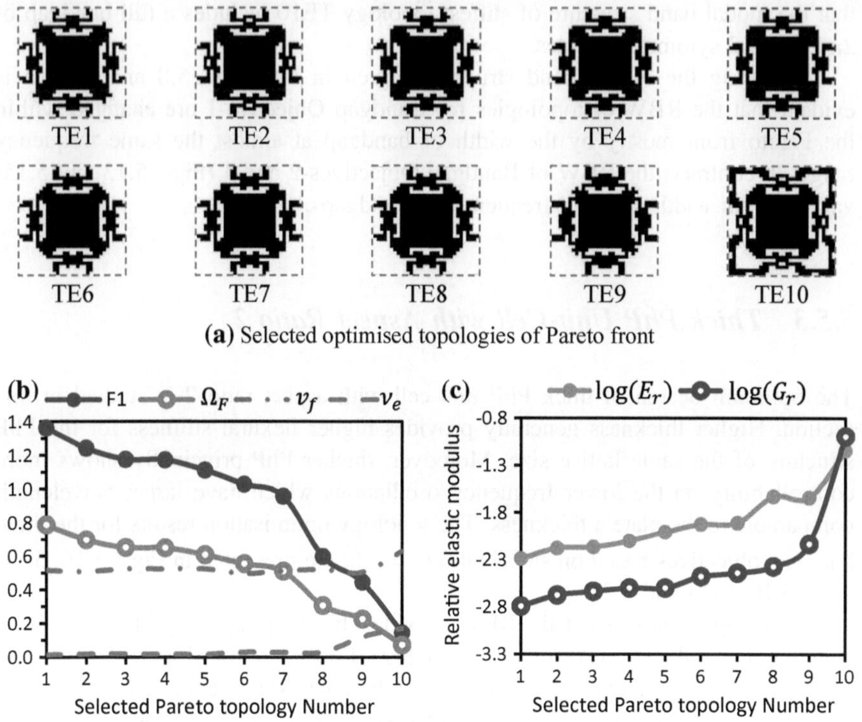

Fig. 5.14 Topology optimisation results of PhP unit-cell with aspect ratio 10 for maximised bandgap objective-3, stress ratio $b_s = \sqrt{2}$; optimised topologies TE1 to TE10 are selected from relevant Pareto front

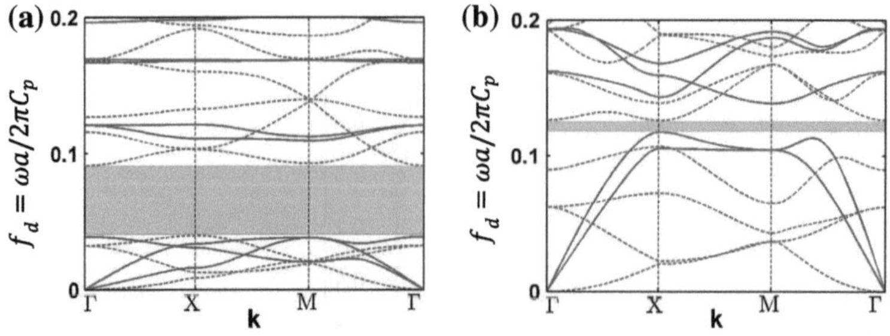

Fig. 5.15 Modal band structure of guided wave modes for **a** topology TE1, and **b** topology TE10 (Fig. 5.14); symmetric and asymmetric modes are shown by solid and dash lines respectively, and complete bandgap is highlighted in grey

which shows a relatively big ramp in both bandgap and stiffness. It is noteworthy that the modal band structure of stiffest topology TE10 includes a full bandgap of 2nd and 3rd symmetric modes.

Comparing the modal band structures given in Figs. 5.6, 5.8 and 5.10 it is evident that the RBW of topologies for Bandgap Objective-1 are changed within the Pareto front mostly by the width of bandgap at almost the same frequency range. In contrary, the RBW of Bandgap Objectives-2 and 3 (Figs. 5.13 and 5.15) vary by both width and mid frequency of bandgaps.

5.5.3 *Thick PhP Unit-Cell with Aspect Ratio 2*

The optimum design of thick PhP unit-cell with aspect ratio 2 is studied in this section. Higher thickness generally provides higher flexural stiffness for the PhP structure of the same lattice size. Moreover, thicker PhP principally shows more controllability on the lower frequency oscillations which have larger wavelength comparable to the plate's thickness. The topology optimisation results for the three bandgap objectives based on stress ratio $b_s = \sqrt{2}$ are presented in Figs. 5.16, 5.17, 5.18, 5.19, 5.20 and 5.21.

For Bandgap Objective-1 the RBW of widest bandgap topology TF1 (Fig. 5.16) is almost twice that of corresponding topology TA1 for thin PhP (Fig. 5.5) around almost the same midgap frequency. Despite considering balanced stress ratio $b = \sqrt{2}$, the shear modulus shows considerably higher gradient compared to elastic modulus. The shear modulus is increased by around 2 orders of magnitude across the Pareto front while the elastic modulus shows slight variation for a large portion of topologies and relatively lower rise at the end (Fig. 5.16c). It is worth noting that the normal stiffness is governed by elastic modulus and Poisson's ratio so can vary by either of these elastic properties. As an example the topology TF9 shows an exceptional mode of stiff topology within the Pareto front which possesses relatively higher elastic modulus but instead lower Poisson's ratio (Fig. 5.16b, c). The

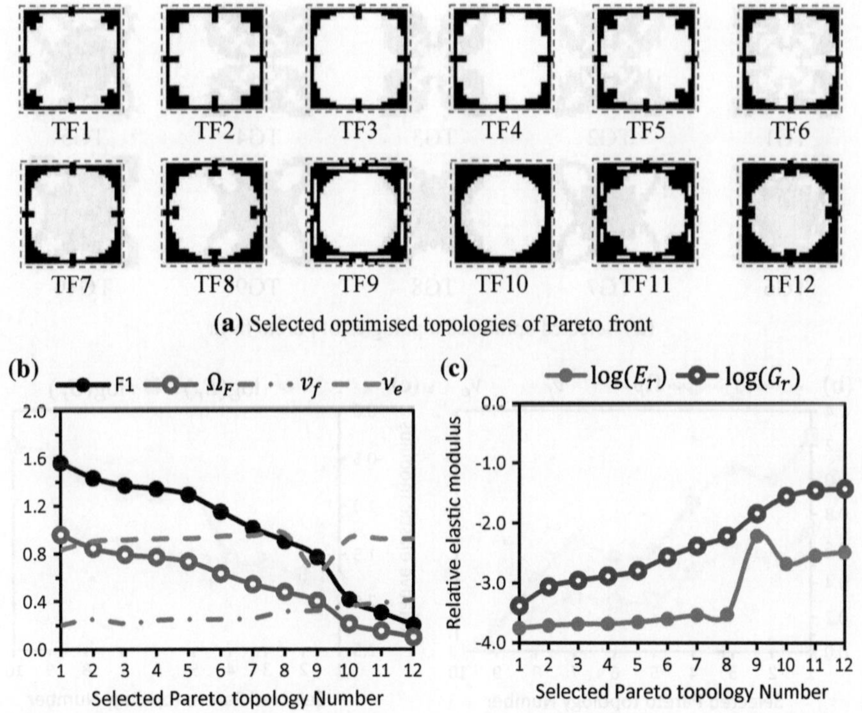

Fig. 5.16 Topology optimisation results of PhP unit-cell with aspect ratio 2 for maximised bandgap objective-1, stress ratio $b_s = \sqrt{2}$; optimised topologies TF1 to TF12 are selected from relevant Pareto front

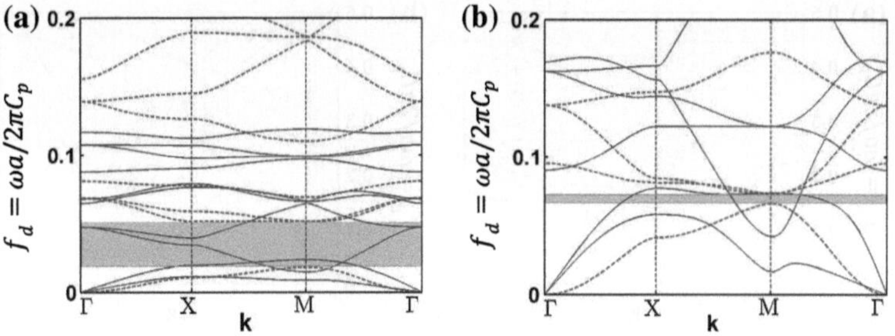

Fig. 5.17 Modal band structure of guided wave modes for **a** topology TF1, and **b** topology TF12 (Fig. 5.16); symmetric and asymmetric modes are shown by solid and dash lines respectively, and lowest bandgap of asymmetric modes is highlighted in grey

modal band structures of extreme topologies TF1 and TF12 presented in Fig. 5.17 show the relevant exclusive bandgaps of the first two asymmetric modal branches.

Regarding the Bandgap Objective-2, dislike thin PhP of 10, the optimisation for lowest possible bandgap between the 2nd and 3rd symmetric modes led to

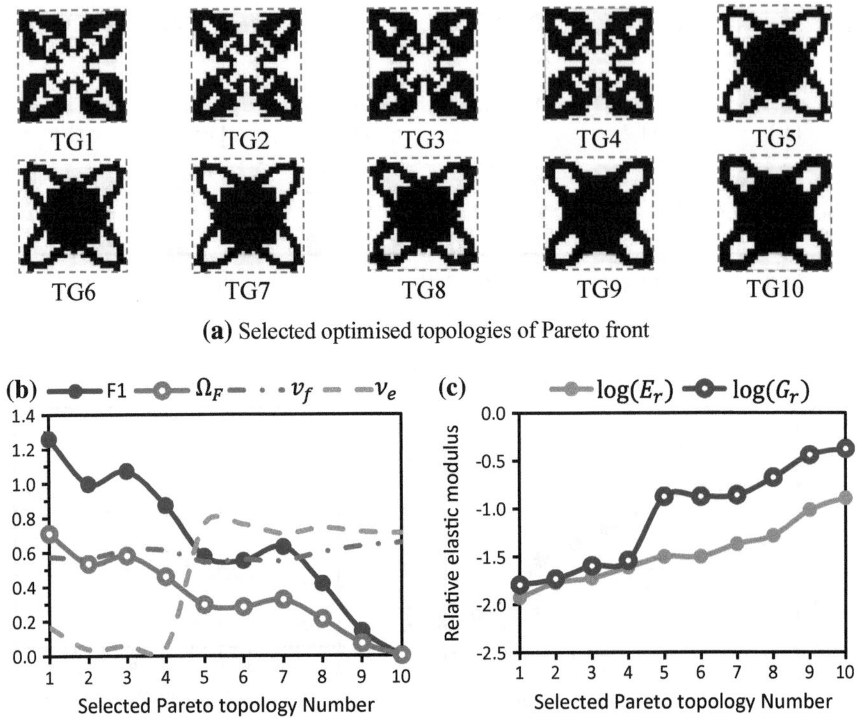

Fig. 5.18 Topology optimisation results of PhP unit-cell with aspect ratio 2 for maximised bandgap objective-2, stress ratio $b_s = \sqrt{2}$; optimised topologies TG1 to TG10 are selected from relevant Pareto front

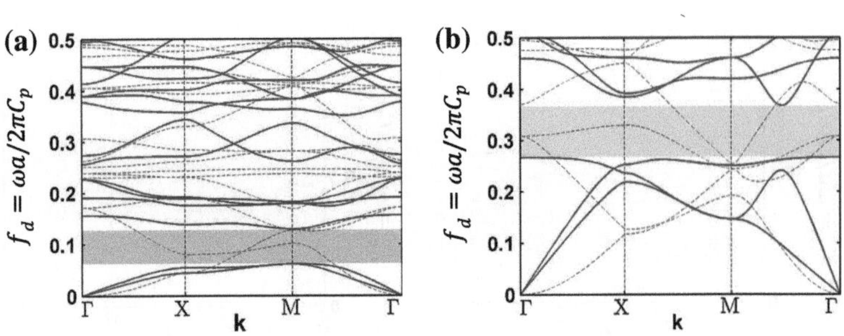

Fig. 5.19 Modal band structure of guided wave modes for **a** topology TG1, and **b** topology TG10 (Fig. 5.18); symmetric and asymmetric modes are shown by solid and dash lines respectively, and lowest bandgap of symmetric modes is highlighted in grey

acceptable topologies with adequate internal connections as demonstrated in Fig. 5.18a. The shear modulus and elastic modulus increase by around 1.5 and 1 orders of magnitude respectively, throughout the Pareto front around the filling fraction 0.6. Shear modulus and Poisson's ratio both show a significant ramp when

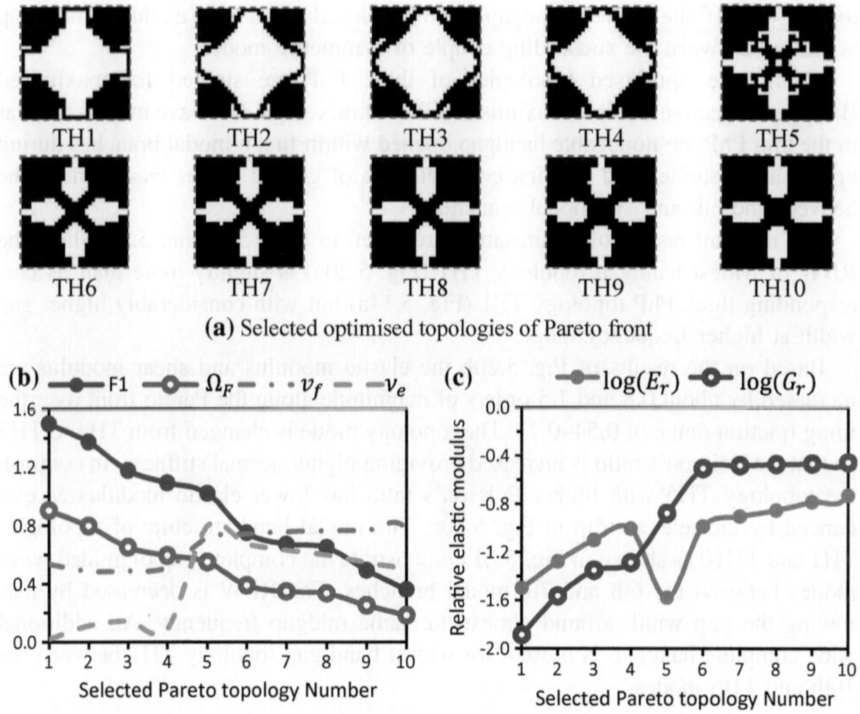

Fig. 5.20 Topology optimisation results of PhP unit-cell with aspect ratio 2 for maximised bandgap objective-3, stress ratio $b_s = \sqrt{2}$; optimised topologies TH1 to TH10 are selected from relevant Pareto front

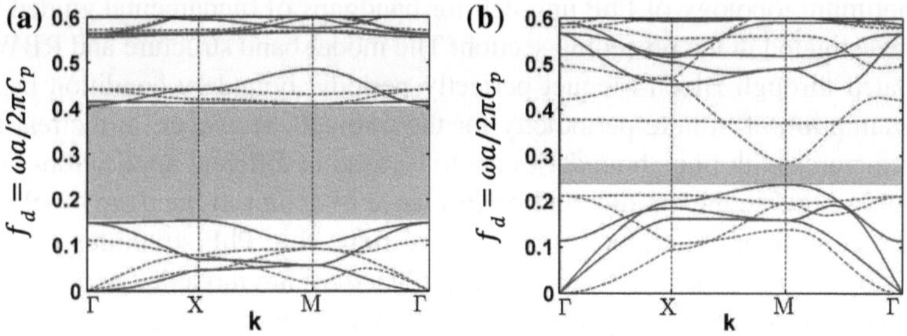

Fig. 5.21 Modal band structure of guided wave modes for **a** topology H-1, and **b** topology H-10 (Fig. 5.20); symmetric and asymmetric modes are shown by solid and dash lines respectively, and lowest complete bandgap is highlighted in grey

the topology changes from TG4 to TG5 with obviously different morphologies leading to higher in-plane stiffness. The modal band structure of widest bandgap topology TG1 and stiffest topology TG10 illustrated in Fig. 5.19 clearly show the relevant exclusive bandgap of desired symmetric modes. Of course, for the stiffest topology TG10 the desired bandgap is almost closed and a wide exclusive bandgap is present between the succeeding couple of symmetric modes.

Finally the optimised topologies of thick PhP are studied for maximised Bandgap Objective-3 to get maximised RBW of mixed guided wave modes. Similar to the thin PhP, no noticeable bandgap opened within first 6 modal branches during optimisation studies and the first complete gap of guided waves was found to be between the 6th and 7th modal branches.

The relevant results of optimisation are given in Figs. 5.20 and 5.21 where the RBW of widest bandgap topology TH1 (Fig. 5.20a) is slightly more than its corresponding thick PhP topology TE1 (Fig. 5.14a) but with considerably higher gap width at higher frequency range.

Based on the results of Fig. 5.20b the elastic modulus and shear modulus are increased by about 0.8 and 1.5 orders of magnitude along the Pareto front over the filing fraction range of 0.54–0.77. The topology mode is changed from TH4 to TH5 so that the Poisson's ratio is increased providing higher normal stiffness. In contrast, the topology TH5 with higher Poisson's ratio has lower elastic modulus as evidenced by the relevant dip in Fig. 5.20c. The modal band structure of topologies TH1 and TH10 as shown in Fig. 5.21 demonstrate the complete gap of guided wave modes between the 6th and 7th modal branches. The RBW is decreased by narrowing the gap width around almost the same midgap frequency. An additional wide complete bandgap is present for widest band gap topology TH1 between the 10th and 11th modes.

5.6 Frequency Response of Finite PhP Structure

The optimum topology of PhP unit-cell for bandgaps of fundamental guided waves was investigated in the preceding section. The modal band structure and RBW were calculated through Bloch-Floquet perfectly periodic boundary condition based on the assumption of infinite periodicity for the unit-cell. However, in the real world, PhP structures with finite boundaries are to be used in different applications and it is essential to realise the validity and performance of achieved bandgap topologies in realistic conditions. Therefore representative finite thin PhP structures of selected Pareto topologies related to the 3 bandgap objectives are modelled and their steady state and transient frequency responses are studied. Intermediate topologies TA6 (Fig. 5.5), TD6 (Fig. 5.12) and TE5 (Fig. 5.14) are chosen from the Pareto fronts of bandgap objectives-1, 2 and 3 respectively, with moderate stiffness and RBW. Lattice periodicity of $a = 10$ mm is assumed leading to plate's thickness $h = 1$ mm for aspect ratio 10.

5.6.1 Modal Band Structure of Selected Topologies

Initially the frequency band structure of selected thin PhP topologies with assumed structural thickness of 1 mm is calculated and the relevant bandgap frequencies are defined as shown in Fig. 5.22. The insets of Fig. 5.22 shows mode shapes of first few modes at arbitrary Brillouin zone point X (i.e. $\mathbf{k} = \{\,\pi/a \quad 0\,\}$) representing wave modes corresponding to the associated modal branches. Accordingly, the lowest bandgap of asymmetric modes for topology TA1 is opened between the fundamental asymmetric Lamb mode A_0 and its resonated branch by Bragg reflection producing bandgap (Fig. 5.22a). Apparently this exclusive gap of asymmetric modes is interrupted by the first symmetric shear horizontal mode SHS_0.

Figure 5.22b shows the exclusive bandgap of symmetric modes for topology TD6 encompassing several asymmetric modes. The lower symmetric modes of this gap are SHS_0 and its folded branch as well as fundamental symmetric Lamb mode S_0. The gap is confined on the upside by relevant resonated modes. The modal band structure of topology TE5 is also given in Fig. 5.22c including the first complete gap of mixed guided waves. The gap is surrounded by fundamental guided wave modes A_0, S_0, SHS_0 and SHA_0 from the downside and their resonated branches from the upside. However, definition of modal branches on upside of complete bandgap is complicated due to their mixed nature.

In order to reveal the nature of resonated modal branches folding back to the first Brillouin zone and confining the maximised bandgap, the modes of bandgap of topology TA6 are further scrutinised in detail. For this purpose the mode shapes corresponding to the points M_1 and M_2 marked in the relevant modal band structure (Fig. 5.22a) are defined and shown in Fig. 5.23. In fact, M_1 and M_2 are the first and the second asymmetric modes at Brillouin zone point X (i.e. $\mathbf{k} = [\,\pi/a \quad 0\,]$) limiting the partial bandgap along the Brillouin zone border ΓX. The first mode M_1 represents a global bending of the lattice uniformly along y-axis, while mode M_2 has also oscillations along y-axis. The mode shapes show the wavelength ζ compared to the unit-cell width a and so the actual wave numbers.

In fact, the second mode M_2 corresponds to the actual wave vector $\mathbf{k} = \{\,\pi/a \quad 2\pi/a\,\}$ which is folded back to the first Brillouin zone due to periodicity of Bloch-Floquet boundary condition (Eq. 3.20). Thus the bandgap of asymmetric wave modes is maximised between fundamental asymmetric Lamb wave mode A_0 and its folded modal branch.

5.6.2 Steady State Frequency Response

An original square plate lattice structure with 20 unit-cells on each side is considered to be excited at the centre (Fig. 5.24a). Hence, the relevant reduced

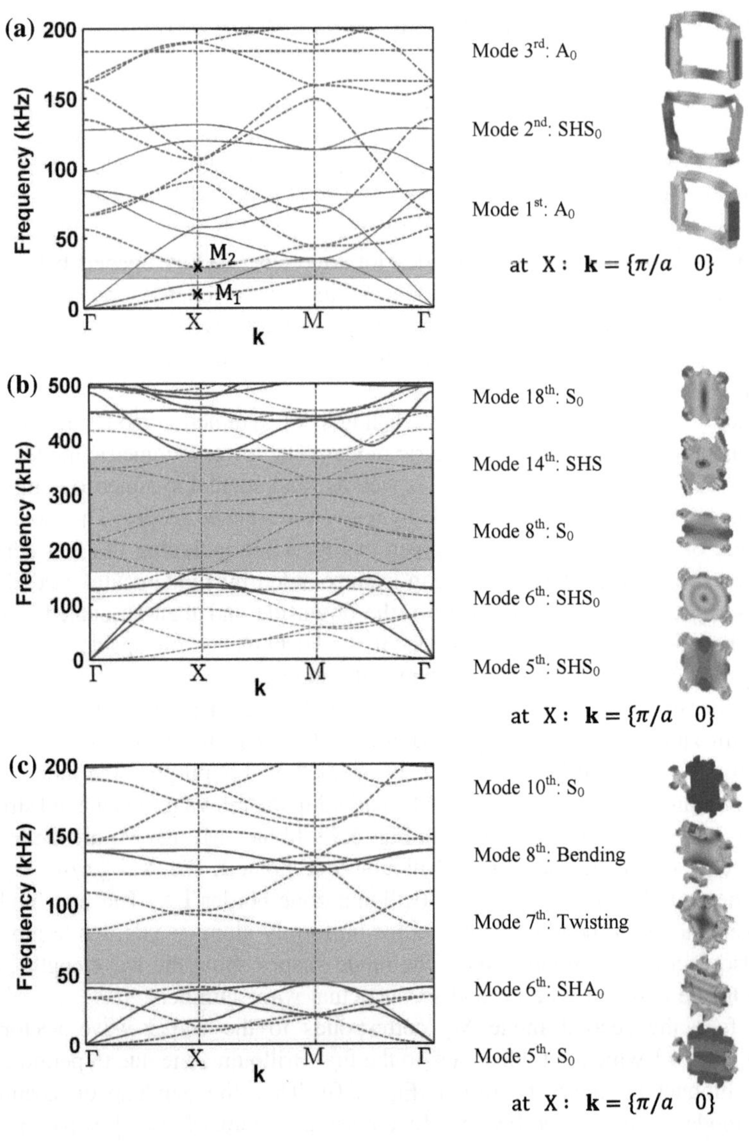

Fig. 5.22 Modal band structure and magnified mode shapes of guided wave modes at Brillouin zone point X for selected topologies **a** TA6, **b** TD6 and **c** TE5; symmetric and asymmetric modes are shown by solid and dash lines respectively, and bandgap highlighted by shaded area

symmetric FEM model of assumed plate structure is developed like the one shown in Fig. 5.24b for selected topology TA6.

Unlike optimisation procedure, no compliant material is considered and the solid part of topology is modelled only. The model is subjected to a probing load at the central point O. Apparently the geometrically central point O of lattice structure of

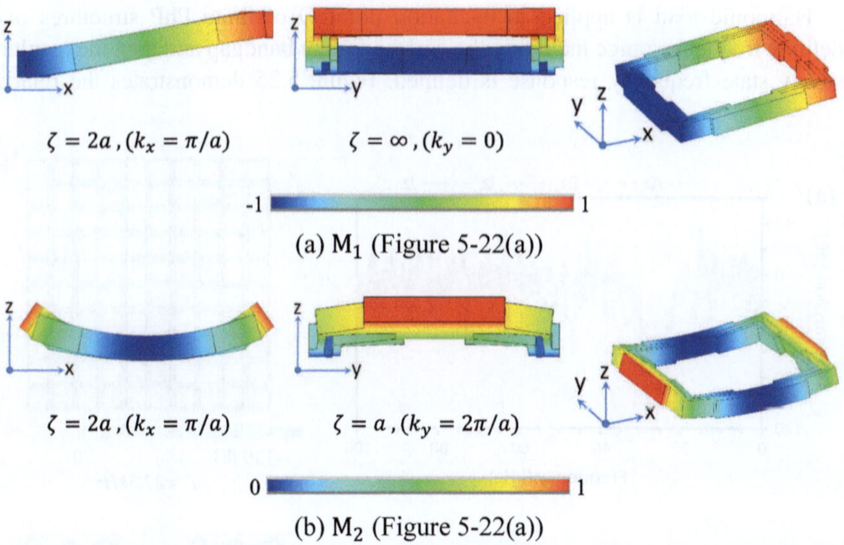

Fig. 5.23 Magnified mode shapes of topology TA6 and normalised contour of transversal displacement w along z-axis corresponding to the **a** first and **b** second asymmetric modal branches at Brillouin zone point X (i.e. $\mathbf{k} = [\pi/a \quad 0]$) marked in the relevant modal band structure (Fig. 5.22a) as M_1 and M_2

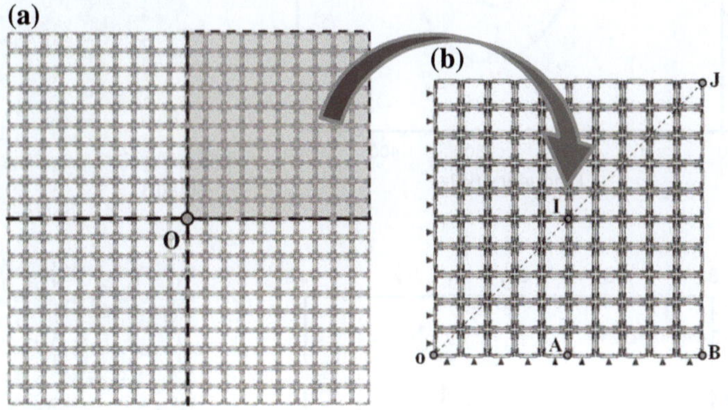

Fig. 5.24 **a** Finite square PhP structure with 20 cells of topology TA6 (Fig. 5.5a), $a/h = 10$ and $h = 1$ mm on each side, and **b** its reduced symmetric model excited at the central point O, to be captured on straight sampling points A and B and diagonal ones I and J

selected topologies is void and so the diagonally nearest solid node is excited instead. Then the transmission of stimulated oscillations is captured at 4 different positions within the plate structure. Two straight points A and B as well as two diagonal points I and J are selected so that the transmission of guided waves to different distances in different angles is determined (Fig. 5.24).

Harmonic load is applied to excitation point O of finite PhP structures over definite frequency range including the corresponding bandgap and then the resultant steady state frequency response is defined. Figure 5.25 demonstrates the relative

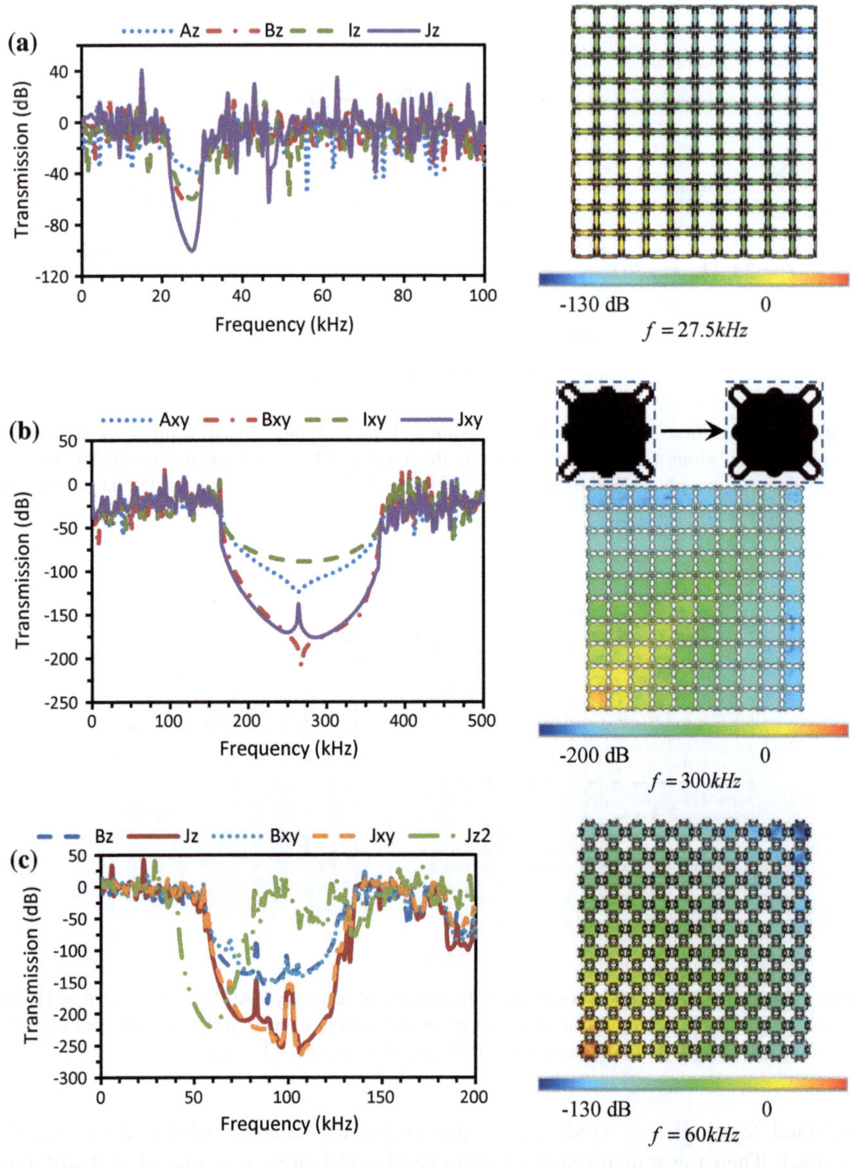

Fig. 5.25 Frequency response of finite PhP structure with selected optimised topologies **a** TA6, **b** refined TD6 and **c** TE5 in logarithmic scale dB; Left: transmission of in-plane (xy) and/or anti-plane (z) polarisations, Right: transmission contour of displacement vector at selected bandgap frequency

transmittance of excited load to specified sampling points A, B, I and J (Fig. 5.24) over the selected frequency range in logarithmic scale dB. In-phase transversal loads (along z-axis) and planar loads (along both x and y axes) are applied to both upper and lower sides of the point O for dominant excitation of asymmetric modes (Fig. 5.25a) and symmetric modes (Fig. 5.25b) respectively.

Only the upper side of excitation point O is stimulated transversally for generation of mixed guided wave modes (Fig. 5.25c). The transmittance of in-plane polarisations (indexed xy) and transversal polarisations (indexed z) are so defined based on the nature of excited wave.

As for PhP structure of asymmetric bandgap topology TA1, the transmission spectrum (Fig. 5.25a) of transversal polarisation is highly attenuated in the same bandgap frequency predicted by the modal band structure of Fig. 5.22a around 22–32 kHz. In order to model the finite structure of symmetric bandgap topology TD6 a polished topology as an inset of Fig. 5.25b is employed. The transmission spectrum of Fig. 5.25b for in-plane polarisation over 0–500 kHz presents the same wide bandgap as shown in the modal band structure of Fig. 5.22b in the frequency range 165–370 kHz. As expected, the lowest and the highest attenuations occur at sampling points A and J respectively, due to their relative distance to the excitation point O.

Lastly, the transmittance of both transversal and in-plane polarisations is determined for the PhP structure of topology TE5 with complete gap of mixed guided waves. Despite the asymmetric nature of aforementioned loading for this case, the related transmission spectrum given in Fig. 5.25b shows dominant symmetric excitation by a wide bandgap corresponding to the calculated gap of symmetric modes (Fig. 5.22c) over around 50–140 kHz.

However, the spectrum of transversal polarisation which principally represents the oscillations of asymmetric modes is resonated at frequency 83 kHz corresponding to presence of resonated A_0 mode at Brillouin zone point Γ (i.e. $\mathbf{k} = [0 \quad 0]$). For more clarity pure asymmetric modes are dominantly excited and the transmission spectrum to point J is calculated (index z2 in Fig. 5.25c). Expectedly the asymmetric bandgap of Lamb mode A_0 in the frequency range 40–80 kHz is evident. Based on the relevant modal band structure of Fig. 5.22c, two asymmetric modes A_0 and SHA_0 interrupt the symmetric bandgap. However, the mode SHA_0 obviously was not dominant in either examined loading cases.

The transmission contour of displacement vector at specified bandgap frequency is also given in each case to show the spatial gradient of frequency response. The contours clearly show the intense decay of oscillations' amplitude all over the structure in all directions through successive number of PhP unit-cells.

5.6.3 Transient Frequency Response

The transient frequency response of finite PhP structure is also defined for asymmetric bandgap topology TA1 (Fig. 5.24) to inspect the attenuation of asymmetric modes as well as passage of symmetric modes within bandgap frequency. Guided wave modes are launched via a 3-cycle sinusoidal tone-burst of bandgap frequency 27.5 kHz modified by Hanning window for high power concentration around this central frequency. The selected central frequency 27.5 kHz corresponds to the dip in Fig. 5.25a. As for asymmetric modes, transversal displacement load w_f with amplitude $w_0 = 0.01$ mm is applied to excitation point O as follows (Hedayatrasa et al. 2014; Veidt and Ng 2011):

$$w_f = (w_0 \sin \omega t)\left(0.5(1 - \cos\frac{\omega t}{N_c})\right) \tag{5.20}$$

for time duration $t = 109\ \mu\text{s}$ corresponding to $N_c = 3$ cycles of frequency 27.5 kHz. The second term of Eq. (5.20) is Hanning window function applied to the sinusoidal signal. Then the transient response of structure is defined up to $t = 1200\ \mu\text{s}$ to let the asymmetric wave modes propagate throughout the plate. Displacement load is preferably applied to the structure to make sure the desired bandgap frequency is excited adequately despite natural resistance of the structure to vibrate within gap frequency range. When force load was examined, the frequency spectrum naturally deviated from the intended bandgap frequency of 27.5–22 kHz due to resonance of A_0 mode confining lower gap frequency at 22 kHz. The time history of normalised amplitude of excited oscillations at point O and transmitted oscillations at points A and I are given in Fig. 5.26a. The initial part of signal concerning loading point O shows the 3-cycle windowed exciting tone-burst. The time-frequency spectrum of the signals is then obtained through Gabor wavelet transform. Wavelet transform is a great tool for time-frequency spectrum decomposition of transient response and among various mother wavelets the Gabor wavelet produces the best resolution in both time and frequency domains (Hedayatrasa et al. 2014; Kishimoto et al. 1995; Ng et al. 2009; Quek et al. 2001). Gabor wavelet is actually a harmonically modulated Gaussian window in the complex form:

$$\psi(t) = \sqrt{4}\pi\sqrt{\frac{2}{\gamma_G}}\text{EXP}\left(-0.5\left(\frac{\omega}{\gamma_G}(t - t_c)\right)^2\right)\text{EXP}(i\omega(t - t_c)) \tag{5.21}$$

where $\gamma_G = \pi\sqrt{2/ln2}$ and t_c is the time at the centre of Gaussian envelope. Figure 5.26b represents the spectrum of excitation point which clearly shows power concentration within bandgap frequency and central frequency 27.5 kHz. However, in spite of intense excitation of bandgap frequency, the spectrum of transmitted oscillations to sampling points A (Fig. 5.26c) and I (Fig. 5.26d) confirms the strong attenuation of waves in the time domain within bandgap frequency range as compared to the relevant band structure in Fig. 5.22a. The logarithmic time-frequency spectrums of transmitted signals to points A and I are also given in Fig. 5.22e, f respectively, to discern the order of magnitude of spectrum. Wave attenuation at point A is obviously higher than point I, despite smaller distance of

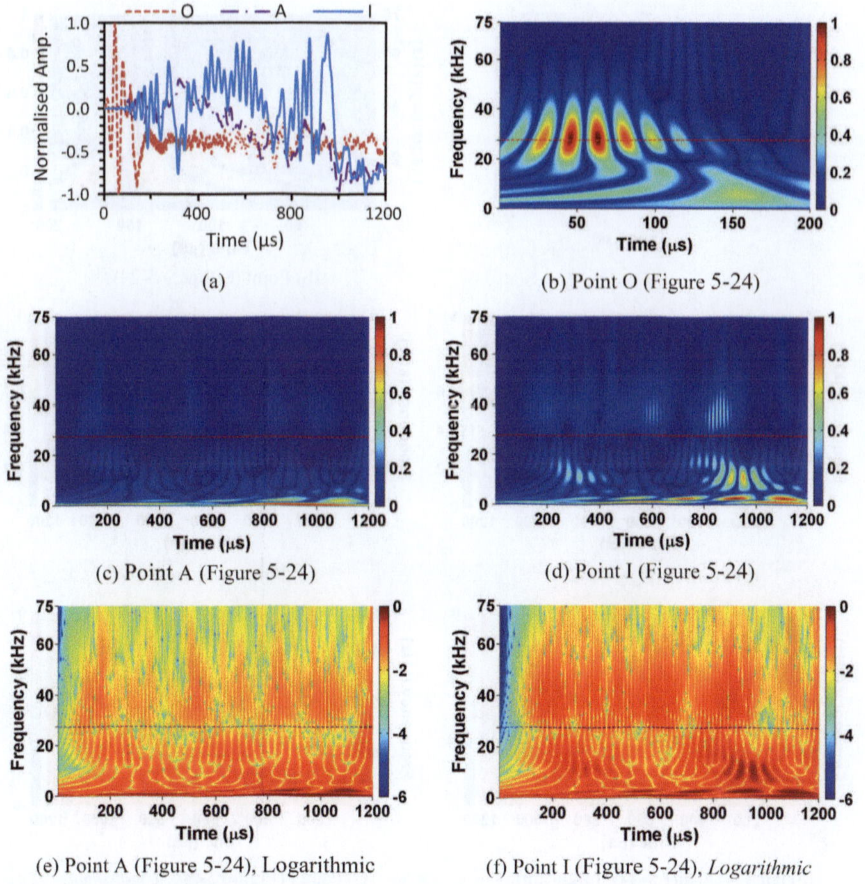

Fig. 5.26 Transient response of the PhP structure of Fig. 5.24 under transversal Hanning window ton-burst at central asymmetric bandgap frequency 27.5 kHz (Fig. 5.25a); time history and Gabor wavelet spectrum of excited and transmitted transversal oscillation w

point A to excitation point. This observation can be justified by wider bandgap width of asymmetric modes along Brillouin border ΓX which corresponds to wave propagation through longitudinal x-axis including point A.

Symmetric guided wave modes are stimulated and analysed further via the same procedure explained for asymmetric modes, but this time by exciting both in-plane polarisations u and v at point O. The time history of the normalised amplitude of resultant excited oscillations at point O and transmitted oscillations at points A and I are given in Fig. 5.27a.

The bandgap frequency 27.5 kHz is properly excited according to the excitation spectrum given in Fig. 5.27b during the excitation time $t = 109\,\mu\text{s}$. The time-frequency spectrums of resultant in-plane oscillations transmitted to points A (Fig. 5.27a, e) and J (Fig. 5.27a, f) represent robust excitation and passage of

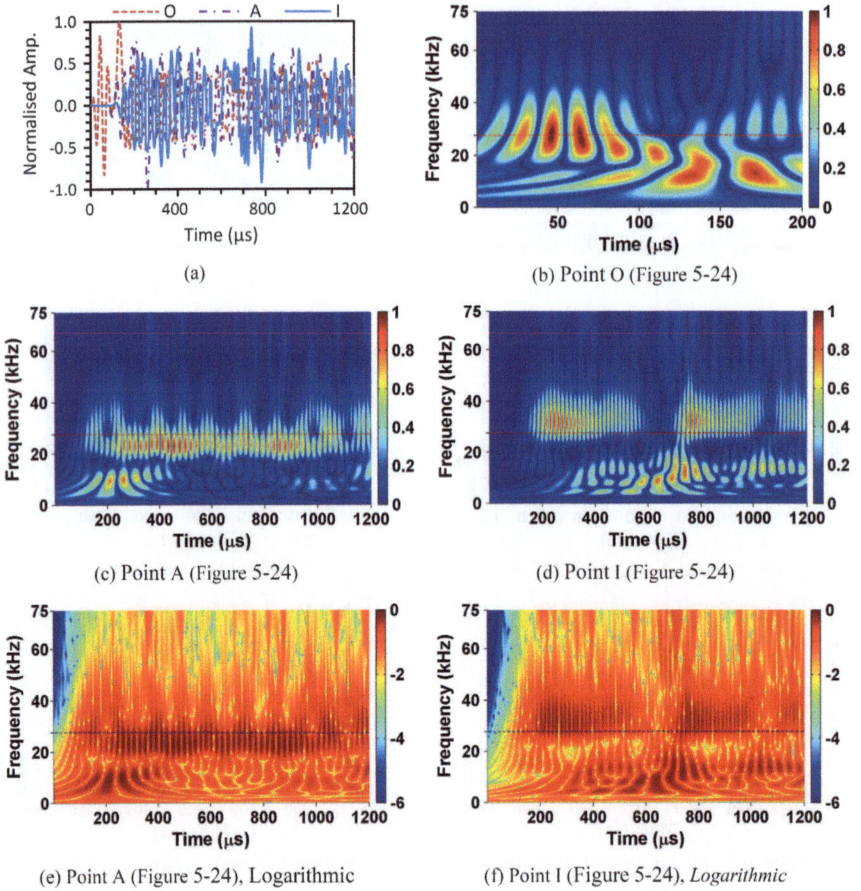

Fig. 5.27 Transient response of the PhP structure of Fig. 5.24 under in-plane Hanning window ton-burst at central asymmetric bandgap frequency 27.5 kHz (Fig. 5.25a); time history and Gabor wavelet spectrum of excited and transmitted in-plane oscillation in xy-plane

symmetric modes throughout the PhP structure within bandgap frequency range 20–30 kHz of asymmetric modes. Despite the power concentration of launched wavelet at bandgap frequency 27.5 kHz, the central frequency of transmitted waves is slightly changed. This is due to existence of partial bandgaps of symmetric modes effective along particular wave vectors, i.e. propagation directions, and resonance of relevant modal branches (Fig. 5.22a). The spectrum of diagonal point I is more noticeably raised to around 35 kHz. This observation can be explained by resonance of two SHS_0 modal branches approaching around frequency 35 kHz corresponding to symmetric Brillouin zone point M (Fig. 5.22a). However, transmitted symmetric waves to both points A and I show high spectrum power inside asymmetric bandgap frequency range 20–30 kHz.

5.7 Concluding Remarks

Evolutionary multi-objective optimisation of porous PhP unit-cell with square symmetry was performed successfully in order to maximise RBW of fundamental guided wave modes with maximal in-plane stiffness where the obtained topology can be produced by perforation of a uniform solid plate. The optimum topology of such single material porous plate structure for widest bandgap naturally favours maximum isolation of scattering segments achievable through weak structural connections. Hence, the strain energy compliance of unit-cell was considered as the second objective to be minimised to get bandgap topologies with maximised stiffness. Complete bandgaps of mixed guided wave modes as well as exclusive bandgaps of symmetric and asymmetric guided wave modes were explored. Polysilicon solid background was considered and very compliant material was assumed to model the void regions of topology for simplified FEM modelling of randomly generated topologies. Finally, frequency response of thin finite PhP structures produced by selected intermediate Pareto topologies of symmetric, asymmetric and complete bandgaps were defined and validity of relevant topologies was confirmed. High attenuation of relative transmission to different sampling points throughout the plate structure was observed within relevant bandgap frequency under steady state harmonic excitation. Moreover, the transient response of finite PhP structure of asymmetric bandgap subjected to windowed tone-burst was calculated. Time-frequency spectrum of transient response obtained via Gabor wavelet transform evidently indicated the attenuation of asymmetric wave modes and passage of symmetric wave modes in the time domain within relevant asymmetric bandgap frequency.

The resultant PhP structures have promising application in selective filtration, excitation, steering and capturing of relevant modes to be used in various areas e.g. structural health monitoring and acoustic energy harvesting.

Following specific conclusions can be drawn from the results of this chapter:

- Taking into account the effective stiffness of porous PhP (and generally PhCr) is essential when optimising for maximised RBW in order to get structurally worthy topologies. This way the Pareto front offers a variety of bandgap topologies between its two extremes. One extreme with maximised bandgap efficiency having lowest stiffness, and the other extreme with maximised stiffness having lowest (and practically very negligible) bandgap efficiency.
- Different in-plane stress ratios were assumed in definition of strain energy compliance in order to achieve dominant shear and normal stiffness. The Pareto fronts and effective elastic properties of optimised topologies proved to be significantly influenced by this prescribed condition as desired.
- Generally a sudden change in the construction of topology appears across the Prato front which offers relatively higher structural stiffness while maintaining bandgap efficiency.

- Optimised topologies for bandgaps of asymmetric wave modes indicated superior bandgap efficiency compared to preceding works available in the literature.
- Thick PhP with aspect ratio 2 generally provides higher RBW compared to thin PhP with aspect ratio 10.
- The filling fraction of topologies on the stiff side of Pareto front is normally higher than those on the compliant side. However, the stiffness does not necessarily increase by filling fraction and predominantly depends on the construction of topology.
- Reducing the solid material properties, i.e. density and elastic modulus, by the order of magnitude of more than 5 to be used as compliant material provides accurate results for modal band analysis of relevant porous unit-cell.

References

Aberg, M., & Gudmundson, P. (1997). The usage of standard finite element codes for computation of dispersion relations in materials with periodic microstructure. *The Journal of the Acoustical Society of America, 102,* 2007.

ANSYS 14.5 element reference.

Bendsoe, M. P., & Sigmund, O. (2003). *Topology optimization (Theory, methods and applications)*. Berlin: Springer.

Berger, H., Kari, S., Gabbert, U., Rodriguez-Ramos, R., Guinovart, R., Otero, J. A., et al. (2005). An analytical and numerical approach for calculating effective material coefficients of piezoelectric fiber composites. *International Journal of Solids and Structures, 42*(21–22), 5692–5714.

Bilal, O. R., & Hussein, M. I. (2012). Topologically evolved phononic material: Breaking the world record in band gap size. In *Photonic and phononic properties of engineered nanostructures* (pp. 826911–826917). International Society for Optics and Photonics.

Charles, C., Bonello, B., & Ganot, F. (2006). Propagation of guided elastic waves in 2D phononic crystals. *Ultrasonics, 44*(0), e1209–e1213.

Chen, J., Chen, L., Chang, C. C., Zhang, Z., Li, W., Swain, M. V., et al. (2016). Micro-CT based modelling for characterising injection-moulded porous titanium implants. *International Journal for Numerical Methods in Biomedical Engineering.*

Dong, H.-W., Su, X.-X., & Wang, Y.-S. (2014). Multi-objective optimization of two-dimensional porous phononic crystals. *Journal of Physics. D. Applied Physics, 47*(15), 155301.

Gonella, S., & Ruzzene, M. (2008). Homogenization and equivalent in-plane properties of two-dimensional periodic lattices. *International Journal of Solids and Structures, 45*(10), 2897–2915.

Halkjær, S., Sigmund, O., & Jensen, J. S. (2006). Maximizing band gaps in plate structures. *Structural and Multidisciplinary Optimization, 32*(4), 263–275.

Hedayatrasa, S., Bui, T. Q., Zhang, C., & Lim, C. W. (2014). Numerical modeling of wave propagation in functionally graded materials using time-domain spectral Chebyshev elements. *Journal of Computational Physics, 258,* 381–404.

Hollister, S. J., & Kikuchi, N. (1992). A comparison of homogenization and standard mechanics analyses for periodic porous composites. *Computational Mechanics, 10*(2), 73–95.

Hussein, M. I., Hamza, K., Hulbert, G. M., Scott, R. A., & Saitou, K. (2006). Multiobjective evolutionary optimization of periodic layered materials for desired wave dispersion characteristics. *Structural and Multidisciplinary Optimization, 31*(1), 60–75.

Kari, S., Berger, H., & Gabbert, U. (2007a). Numerical evaluation of effective material properties of randomly distributed short cylindrical fibre composites. *Computational Materials Science, 39*(1), 198–204.

Kari, S., Berger, H., Rodriguez-Ramos, R., & Gabbert, U. (2007b). Computational evaluation of effective material properties of composites reinforced by randomly distributed spherical particles. *Composite Structures, 77*(2), 223–231.

Kaw, A. K. (2005). *Mechanics of composite materials*. Boca Raton: CRC press.

Khelif, A., Aoubiza, B., Mohammadi, S., Adibi, A., & Laude, V. (2006). Complete band gaps in two-dimensional phononic crystal slabs. *Physical Review E, 74*(4), 046610.

Kishimoto, K., Inoue, H., Hamada, M., & Shibuya, T. (1995). Time frequency analysis of dispersive waves by means of wavelet transform. *Journal of Applied Mechanics, 62*(4), 841–846.

Krushynska, A. O., Kouznetsova, V. G., & Geers, M. G. D. (2014). Towards optimal design of locally resonant acoustic metamaterials. *Journal of the Mechanics and Physics of Solids, 71*, 179–196.

Kundu, T. (2004). *Ultrasonic nondestructive evaluation—Engineering and biological material characterization*. (Boca Raton: CRC PRESS).

Li, J., Wang, Y.-S., & Zhang, C. (2009). Finite element method for analysis of band structures of phononic crystal slabs with Archimedean-like Tilings. *IEEE International Ultrasonics Symposium*, 1548–1551.

Manktelow, K. L., Leamy, M. J., & Ruzzene, M. (2013). Topology design and optimization of nonlinear periodic materials. *Journal of the Mechanics and Physics of Solids, 61*(12), 2433–2453.

Ng, C.-T., Veidt, M., & Rajic, N. (2009). Integrated piezoceramic transducers for imaging damage in composite laminates. In *Second International Conference on Smart Materials and Nanotechnology in Engineering* (pp. 74932M–74932M–74938). International Society for Optics and Photonics.

Olhoff, N., Niu, B., & Cheng, G. (2012). Optimum design of band-gap beam structures. *International Journal of Solids and Structures, 49*(22), 3158–3169.

Oliveira, J., Pinho-da-Cruz, J., & Teixeira-Dias, F. (2009). Asymptotic homogenisation in linear elasticity. Part II: Finite element procedures and multiscale applications. *Computational Materials Science, 45*(4), 1081–1096.

Olsson Iii, R. H., & El-Kady, I. F. (2009). Microfabricated phononic crystal devices and applications. *Measurement Science & Technology, 20*(1), 012002.

Pennec, Y., Vasseur, J. O., Djafari-Rouhani, B., Dobrzyński, L., & Deymier, P. A. (2010). Two-dimensional phononic crystals: Examples and applications. *Surface Science Reports, 65* (8), 229–291.

Pinho-da-Cruz, J., Oliveira, J., & Teixeira-Dias, F. (2009). Asymptotic homogenisation in linear elasticity. Part I: Mathematical formulation and finite element modelling. *Computational Materials Science, 45*(4), 1073–1080.

Pratap, A., Agarwal, S., & Meyarivan, T. (2002). A fast and elitist multiobjective genetic algorithm: NSGA-II. *IEEE Transactions on Evolutionary Computation, 6*(2), 182–197.

Quek, S., Wang, Q., Zhang, L., & Ong, K. (2001). Practical issues in the detection of damage in beams using wavelets. *Smart Materials and Structures, 10*(5), 1009.

Rupp, C. J., Evgrafov, A., Maute, K., & Dunn, M. (2007). Design of phononic materials/structures for surface wave devices using topology optimization. *Structural and Multidisciplinary Optimization, 34*(2), 111–121.

Sharpe, W. N., Yuan, B., Vaidyanathan, R., & Edwards, R. L. (1997). Measurements of Young's modulus, Poisson's ratio, and tensile strength of polysilicon. In *Proceedings of Tenth Annual International Workshop on Micro Electro Mechanical Systems, 1997. MEMS'97, IEEE* (pp. 424–429), IEEE.

Steven, G. (2006). Homogenization and inverse homogenization for 3D composites of complex architecture. *Engineering Computations, 23*(4), 432–450.

Timoshenko, S. P., & Woinowsky-Krieger, S. (1959). *Theory of plates and shells*. New York: McGraw-hill.

Veidt, M., & Ng, C.-T. (2011). Influence of stacking sequence on scattering characteristics of the fundamental anti-symmetric Lamb wave at through holes in composite laminates. *The Journal of the Acoustical Society of America, 129*(3), 1280–1287.

Wu, T.-T., Hsu, J.-C., & Sun, J.-H. (2011). Phononic plate waves. *IEEE Transactions on Ultrasonics, Ferroelectrics and Frequency Control, 58*(10), 2146–2161.

Xia, Z., Zhang, Y., & Ellyin, F. (2003). A unified periodical boundary conditions for representative volume elements of composites and applications. *International Journal of Solids and Structures, 40*(8), 1907–1921.

Chapter 6
Optimisation of Porous 2D PhPs: Topology Refinement Study and Other Aspect Ratios

6.1 Introduction

The contribution of effective structural stiffness in evolution and optimality of porous 2D PhPs with maximised RBW was studied in Chap. 5. Square symmetric topology with resolution of 32×32 was considered and structurally worthy porous PhPs were achieved. Higher topology resolutions leads to increased versatility of design domain and potentially can introduce much more effective topologies. However, the design space becomes larger and increases the computational intensity and convergence rate of implemented stochastic optimisation algorithm. Although the effective structural stiffness can be enhanced micromechanically, tiny delicate features may appear in the topology introducing two major limitations: (i) manufacturability, and (ii) local stress intensity and vulnerability.

It is well known that an appropriate projection approach can capably handle the aforementioned limitations by introducing a design variable space to be mapped into topology resolution and further relevant FEM model. In this way the topology resolution is increased while limiting the minimum feature size and design space size. In this chapter a topology refinement study is performed first by doubling the topology resolution ($32 \times 32 \rightarrow 64 \times 64$). The optimisation algorithm is also improved for higher computational efficiency and controlled evolution of topologies in the last stages of optimisation. The coarse 32×32 topology with aspect ratio 10 is first obtained through improved algorithm and afterwards is refined to 64×64 at the secondary stage. The Pareto fronts are presented and analysed and relative performance of achieved topologies is evaluated to explore the efficiency of topology refinement.

Other intermediate aspect ratios of 4 and 6 are also optimised and the distinction and bandgap-stiffness efficiency of topologies (for the same unit-cell width and different aspect ratios) are investigated. Bandgap Objective-1 and Bandgap Objective-3, as introduced in Chap. 5, are investigated for bandgaps of asymmetric modes and complete bandgap of guided wave modes. The results will be used for production of PhP test specimens and experimental evaluation of optimised topologies.

S. Hedayatrasa, *Design Optimisation and Validation of Phononic Crystal Plates for Manipulation of Elastodynamic Guided Waves*, Springer Theses,
https://doi.org/10.1007/978-3-319-72959-6_6

6.2 Modified Topology Optimisation Strategy

Only one-eighth of the PhP unit cell is considered as highlighted in Fig. 6.1a for the independent design domain of square symmetric solutions. To further reduce the number of independent design variables to which stochastic search algorithms are particularly sensitive, we use the reduced dimension approach of Guest and Smith Genut (2010). The approach is based on the projection method, which controls the minimum length scale of features by projecting a constitutive material from a design variable onto finite element centroids in a radial manner over the prescribed length scale (Guest et al. 2004).

In this study nodal design variables are defined with a small projection radius of one element size so that any design variable fills its 4 surrounding elements. A low resolution topology of 32 × 32 with independent design variable size of 136 is initially introduced as shown in Fig. 6.1a with design variables on element centroids (Fig. 6.1b).

After substantial optimisation of topology with this reduced design variable space, nodal design variables are assigned to the domain (without immediate effect to the topology) creating refined design variable space of size 496 relevant to the refined resolution of 64 × 64 partially shown in Fig. 6.1c.

The boundary nodes are preferably excluded from design variable field in order to limit the radius of orthogonal interconnecting elements of topology (darker elements in Fig. 6.1) when performing topology refinement. Thinner interconnections immediately introduce topologies with generally much lower stiffness and higher bandgap efficiency. Hence, the refined nodal design variable space shown in Fig. 6.1c maintains the focus of Pareto solutions without introducing such major changes in the design space.

As discussed earlier in Chap. 5, usually a compliant material needs to be considered as void to maintain the continuity of design domain for gradient based optimisation of porous media. Even in a stochastic optimisation, like GA, this

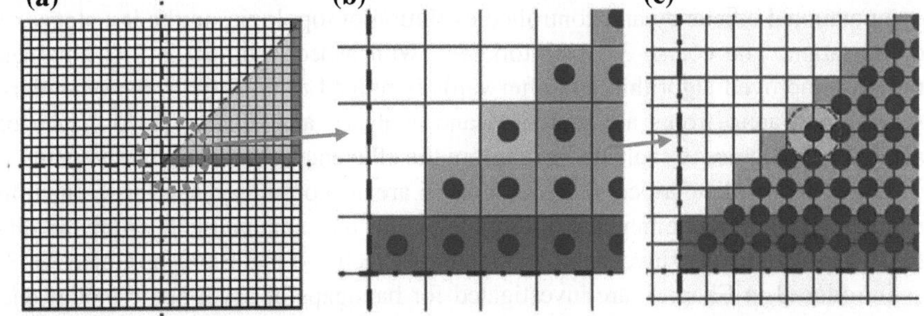

Fig. 6.1 **a** Square symmetric topology with coarse resolution of 32 × 32, **b** close view of relevant design variables on element centroids, and **c** nodal design variables assigned to the refined topology resolution of 64 × 64 to be projected into 4 adjacent elements

technique is preferred. This is to avoid complexities associated with sole modelling of the solid region of randomly generated topology and adopting the relevant boundary conditions, through a fast and efficient approach.

In this part of research the void section of any candidate topology and relevant boundary conditions are removed before analysis by taking the advantage of ANSYS APDL features. This approach significantly increases the computational performance of multi-objective fitness evaluation. GA operations are applied in a dynamic multistage manner. Due to computationally intense nature of defined optimisation problem it is very difficult to run many trial optimisations and optimise the GA settings. However, based on the experience from earlier optimisation studies introduced in previous chapters, the following settings proved to be efficient. According to Table 6.1, the topology optimisation is mainly performed in Stage-1 for reduced design variable space for 120 generations followed by finishing optimisation Stage-2 of 30 generations. In this finishing stage a random *edge mutation* is exclusively applied to the design variables corresponding to the interfacial pixels of topology.

Although the mutation exploits the design space for new designs, it may introduce undesirable checkerboards. Hence a repair operation is also applied at this finishing stage with the same probability of segment elimination. This repair operation targets randomly selected isolated void pixels of topology (not randomly selected arbitrary pixel which is not necessarily void and isolated). In this way the optimised Pareto front is more likely to converge to topologies with smooth boundaries and minimised checkerboards, unless some features persist in order to maintain the elitism of solution set.

When transferring to refinement stage, the design space size significantly increases from 136 independent design variables to 496. The interior regions of the solid and void area are filled with many redundant design variables with no effect on the topology. This is due to overlap of projected areas by adjacent design variables in refined design space (Fig. 6.1c) (Guest and Smith Genut 2010). However no major change in topology is expected and the objective is to just refine the boundaries of Pareto topologies of preceding stage. Consequently, the population size is maintained the same, crossover probability is reduced to 0.5 and like Stage 2 the mutation is predominantly performed on the design variables pertaining to the edge of topology. This Edge Mutation is varied from 0.1 to 0.01 throughout any stage to reduce biased disturbance of GA random search.

Table 6.1 GA settings used in topology refinement study of 2D porous PhP

Stage	Population size	Design space size	Probability				Generations
			Crossover	Mutation	Edge mutation	Segment elimination	
1	200	136	0.9	0.3–0.02	–	0–0.5	120
2	200	136	0.9	0.01–0	0.1–0.01	0.5–1	30
3	200	496	0.9	0.01–0	0.1–0.01	0.5–1	50

GA population size of 200 is chosen for all stages of optimisation. In Stage 1 the design space is initially searched by high mutation probability 0.3 for only 5 generations to possibly have a good resource of initiating population. Then mutation probability is progressively decreased for finer search during optimisation. Segment elimination probability is also added and gradually increased to avoid disturbance of genomes in the early stages of optimisation. The same Bandgap Objectives already introduced in Chap. 5 are considered and stress ratio $b_s = \sqrt{2}$ [Eq. (5.16)] is assumed for definition of stiffness objective for equal contribution of shear and normal terms.

6.3 Bandgaps of Asymmetric Guided Wave Modes (Bandgap Objective-1)

In this section the results concerning optimisation of coarse and refined 2D PhPs for bandgaps of asymmetric modes through improved optimisation strategy are presented. Bandgap Objective-1 is taken for lowest bandgap between first couple of asymmetric modal branches. From 200 dissimilar topologies achieved in Pareto fronts, optimised topologies TI1-TI8 (with coarse resolution 32 × 32) and TJ1-TJ8 (with refined resolution 64 × 64) and their relative location on the relevant Pareto fronts are presented in Fig. 6.2. The topologies are selected such that the extreme topology with maximised RBW (TJ1 and TI1) and a variety of other topologies across Pareto front are included.

By comparison of Pareto fronts shown in Fig. 6.2c it is evident that the topology refinement significantly enhances the bandgap efficiency and structural stiffness of optimised solutions, and higher RBW is achieved through higher stiffness. The small bold dots in Fig. 6.2c show all obtained individual Pareto topologies and present the spread of solutions across Pareto Front. As expected, due to versatility provided by higher resolution, the refined topologies are almost uniformly distributed over the Pareto front in contrast to the coarse topologies which appear in a few clusters.

A wider spread of topologies are obtained on the stiff side through refined topology. However, there is a considerable gap among the solutions of refined topology between topologies TJ4 and TJ5. The sudden change in the construction of these topologies shown in Fig. 6.2b explains a shift in the topology mode which provides slightly lower bandgap efficiency at much higher stiffness. A similar topology gradient is also observable among the optimised coarse topologies (Fig. 6.2a) between TI4 and TI5. Of course the gap between these two coarse topologies is not much, compared with the difference of TJ4 and TJ5 (Fig. 6.2c). The distinct structural properties of these topologies can be further evaluated by comparing their elastic properties.

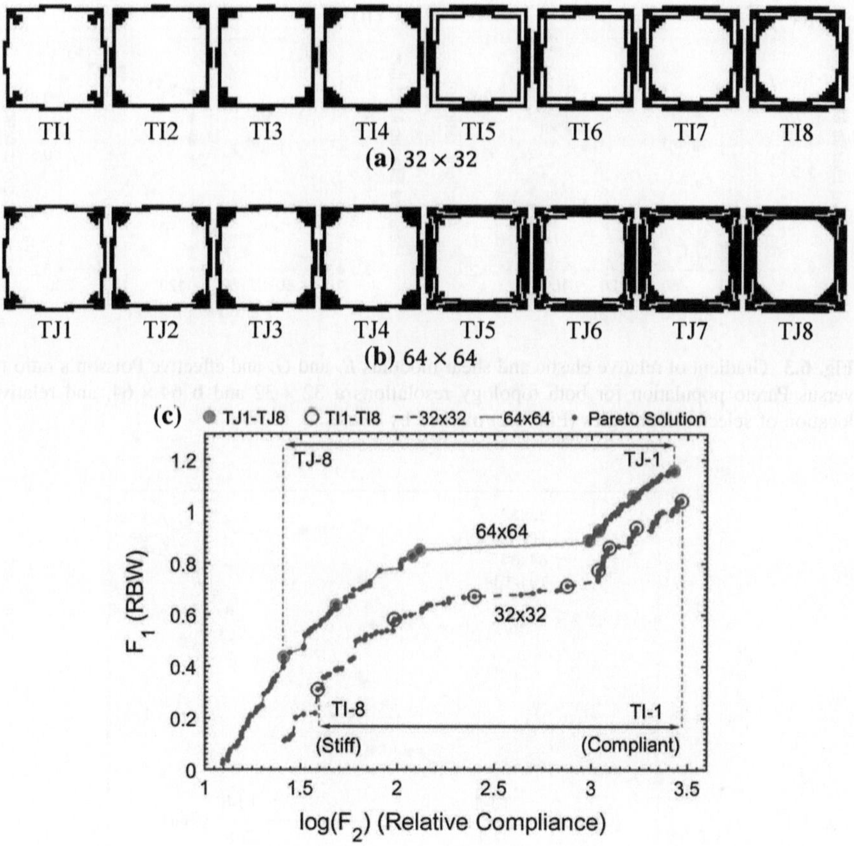

Fig. 6.2 Selected Pareto optimised topologies with resolution **a** 32 × 32 and **b** 64 × 64, and **c** relevant Pareto fronts for Bandgap Objective-1, stress sate ratio $b_s = \sqrt{2}$ and aspect ratio 10

The variation of relative elastic modulus E_r and G_r and effective Poisson's ratio ν_e across the achieved Pareto population is shown in Fig. 6.3a, b for topology resolutions 32 × 32 and 64 × 64 respectively.

The relative elastic and shear modulus is gradually increased through degradation of bandgap efficiency (from left to right) and a sudden jump appears corresponding to the aforementioned gap in the relevant Pareto fronts. The change occurred in the topology mode also causes a drop in its effective Poisson's ratio. The gradient of optimised bandgap frequency range across the Pareto fronts is then shown in Fig. 6.4 and bandgap efficiency of relevant coarse and refined topologies are compared. The selected Pareto topologies of Fig. 6.2a, b are also marked to signify their relative bandgap frequency range. It is noteworthy that the horizontal axis of Fig. 6.4 is just Pareto optimised topology number and does not show the corresponding coarse and refined topologies. The smoother gradient and generally wider bandgaps of refined topologies and the discontinuity in bandgap range of

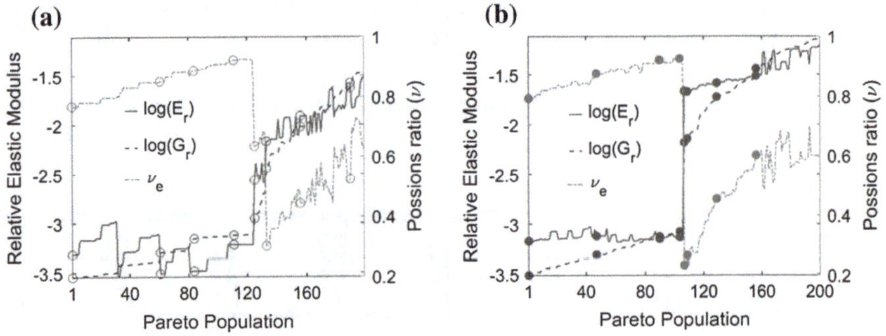

Fig. 6.3 Gradient of relative elastic and shear modulus E_r and G_r and effective Poisson's ratio ν_e versus Pareto population for both topology resolutions **a** 32 × 32 and **b** 64 × 64, and relative location of selected topologies (Fig. 6.2) marked by circles

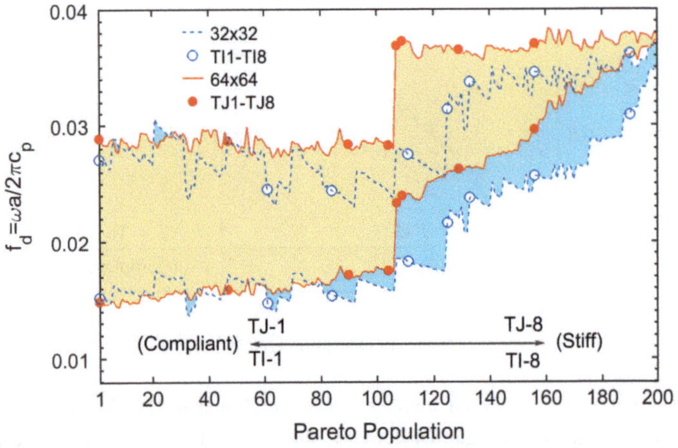

Fig. 6.4 Gradient of bandgap frequency range versus Pareto optimised population of both topology resolutions 32 × 32 and 64 × 64 (Fig. 6.2), and relative location of selected topologies marked by circles

refined topologies is evident. For example, the bandgap of the stiffest refined topology TJ8 starts at lower frequency and ends at higher frequency compared with that of stiffest coarse topology TI8.

Other aspect ratios 4 and 6 are also optimised for Bandgap Objective-1 in order to compare their relative efficiency compared with aspect ratio 10. Selected Pareto topologies and their location in the relevant Pareto fronts are shown in Fig. 6.5, and Pareto front of other aspect ratio 10 from Fig. 6.2c is repeated for comparison. The horizontal axis is the relative in-plane compliance of topologies with respect to pure solid material. So, it should be noted that overall stiffness of topologies (including flexural stiffness) is dependent on the aspect ratio as well which should be taken into account when comparing relative performance of solutions.

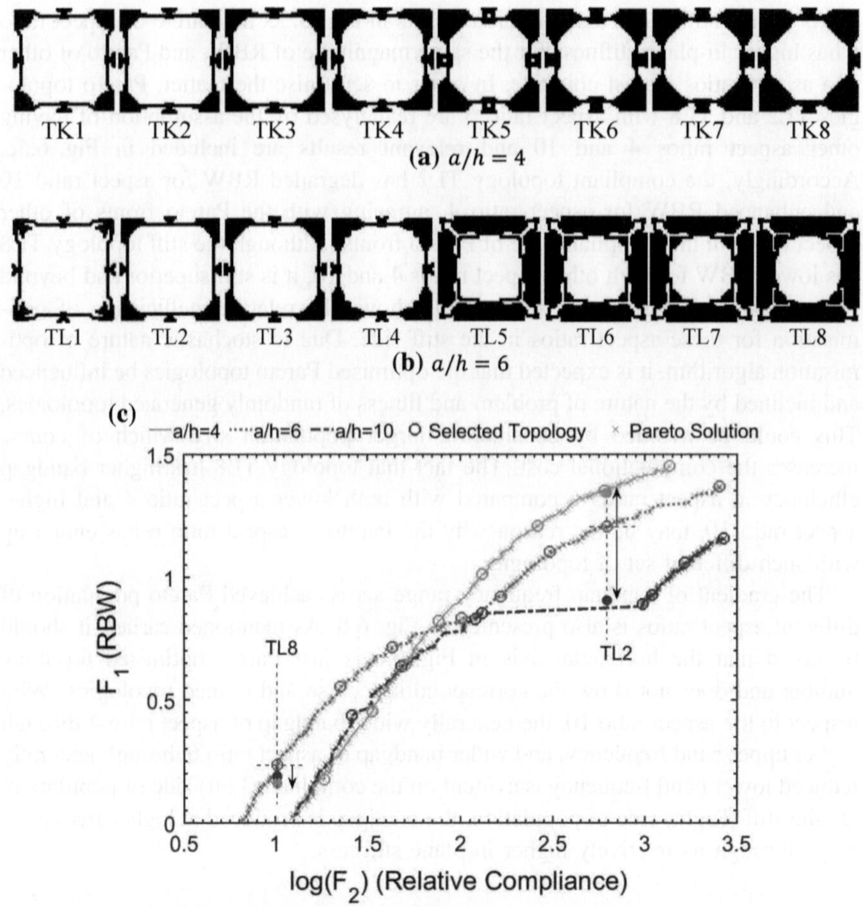

Fig. 6.5 Selected Pareto optimised topologies with resolution 64 × 64 and aspect ratios **a** 4 and **b** 10, and **c** relevant Pareto fronts for Bandgap Objective-1 and stress sate ratio $b_s = \sqrt{2}$ including the Pareto of aspect ratio 10 for comparison (Fig. 6.2c)

The smooth variation of optimised topologies for aspect ratio 4 (Fig. 6.5a) is evident in contrast to the aspect ratio 6 with a sudden variation from topology TL5 to TL5 (Fig. 6.5b). Although optimised topology of aspect ratio 4 has such significant change like aspect ratio 10 (Fig. 6.2b), it just causes a change in the slope of relevant Pareto front and not a big gap in Pareto solutions like that of aspect ratio 10 (Fig. 6.5c). According to the Pareto fronts presented in Fig. 6.5c, for the same unit-cell width, lower aspect ratio leads to higher RBW in the compliant (Right) side.

It is obvious that lower aspect ratio of the same unit-cell width leads to thicker PhP with higher relative flexural stiffness. Thus, lower aspect ratio, were applicable, in general has superior bandgap and structural efficiency compared with higher aspect ratio, for specified lattice periodicity.

As for the stiff (Left) side of Pareto fronts in Fig. 6.5c, the Pareto of aspect ratio 6 has higher in-plane stiffness for the same magnitude of RBW, and Pareto of other two aspect ratios almost coincide. In order to scrutinise the matter, Pareto topologies TL2 and TL8 with aspect ratio 6 are reanalysed by the assumption of having other aspect ratios 4 and 10 and relevant results are included in Fig. 6.5c. Accordingly, the compliant topology TL2 has degraded RBW for aspect ratio 10 and enhanced RBW for aspect ratio 4, agreeing with the Pareto fronts of other aspect ratios in the compliant side of Pareto fronts. Although the stiff topology TL8 has lower RBW for both other aspect ratios 4 and 10, it is still superior and beyond Pareto fronts of these aspect ratios. This shows the relative inefficiency of optimisation for these aspect ratios in the stiff side. Due to stochastic nature of optimisation algorithm, it is expected that the optimised Pareto topologies be influenced and inclined by the nature of problem and fitness of randomly generated topologies. This could be avoided by considering larger population size which of course increases the computational cost. The fact that topology TL8 has higher bandgap efficiency at aspect ratios 6 compared with both lower aspect ratio 4 and higher aspect ratio 10, may be the reason why the Pareto of aspect ratio 6 has ended up with such different set of topologies.

The gradient of bandgap frequency range across achieved Pareto population of different aspect ratios is also presented in Fig. 6.6. As mentioned earlier, it should be noted that the horizontal axis in Fig. 6.6 is just Pareto optimised topology number and does not show the corresponding coarse and refined topologies. With respect to the aspect ratio 10, the generally wider bandgap of aspect ratio 4 through higher upper band frequency, and wider bandgap of aspect ratio 6 through generally reduced lower band frequency is evident on the compliant (Left) side of population. On the stiff (Right) side of population, the bandgap is produced at higher frequency range through its relatively higher in-plane stiffness.

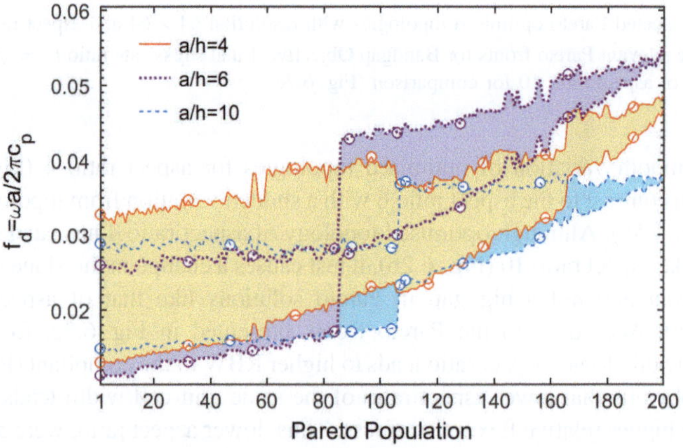

Fig. 6.6 Gradient of bandgap frequency range versus Pareto optimised population for aspect ratios 4, 6 and 10, and relative location of selected topologies marked by circles

6.4 Complete Bandgaps of Mixed Guided Waves Modes (Bandgap Objective-3)

Following Sect. 7.2 concerning bandgaps of asymmetric wave modes, in this section topology refinement study is performed for complete bandgaps of guided waves. Bandgap Objective-3 is taken, as introduced in Chap. 5, to maximise RBW between 6th and 7th modal branches. Selected optimised coarse and refined topologies are shown in Fig. 6.7a, b respectively, and the relevant Pareto fronts are presented in Fig. 6.7c. Just like the results of Sect. 7.2 (Fig. 6.2), the 164 refined Pareto topologies are uniformly distributed over a smooth Pareto front compared with fewer 136 coarse topologies which appear in a few clusters. However, in contrary to the bandgaps of asymmetric wave modes (Fig. 6.2), the topology refinement of complete bandgaps does not lead to major improvement in shifting the Pareto front towards more efficient topologies. This may be due to having a stricter objective to open the bandgap within all types of modal branches instead of specific (symmetric or asymmetric) type. This restriction is also evident from fewer number of dissimilar Pareto topologies achieved for complete bandgaps from initial population size of 200, compared with 200 Pareto topologies obtained for bandgaps of asymmetric modes.

The optimised solutions achieved on the extreme stiff side of Pareto front (Fig. 6.7c) have practically no bandgap efficiency with negative RBW. These topologies on the stiff side of Pareto have relatively narrow complete bandgaps which are actually overestimated due to rough modelling performed during optimisation process for the sake of computational intensity. The modal band analysis is performed over just 5 points on each border of Brillouin zone during optimisation with just one layer of solid-shell elements through the thickness. But the achieved Pareto solutions are analysed over 25 points on each Brillouin zone border and are modelled by 4 element layers to appropriately model the tiny features of topology. Pareto of refined topology have fewer such invalid topologies naturally because a finer in-plane mesh is considered during optimisation of these topologies.

According to the Fig. 6.7c, similar to the Pareto of asymmetric bandgaps (Fig. 6.2c), a gap in distribution of solutions appears in the intermediate section of Pareto front and persists during the refinement stage. This discontinuity of Pareto front is due to a sudden change in the topology mode leading to much higher overall stiffness while maintaining the bandgap efficiency.

The distinction of extreme topologies related to this section of Pareto front is evident as shown in Fig. 6.7a, b: coarse topologies TM6 and TM7, and refined topologies TN5 and TN6. This matter is clearly shown further through scrutinising the variation of elastic properties and bandgap frequency range throughout Pareto fronts.

The gradient of relative elastic modulus E_r and G_r and effective Poisson's ratio ν_e across the achieved Pareto population is shown in Fig. 6.8a, b for topology resolutions 32 × 32 and 64 × 64 respectively. Accordingly, the relative elastic and shear modulus is gradually increased through degradation of bandgap efficiency

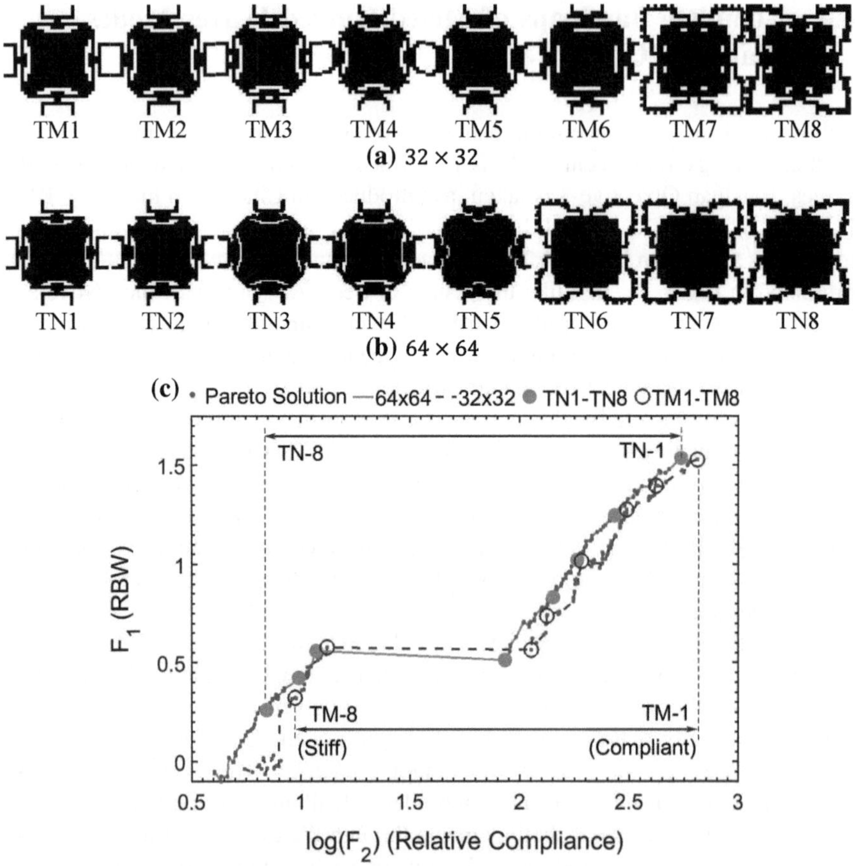

Fig. 6.7 Selected Pareto optimised topologies with resolution **a** 32 × 32 and **b** 64 × 64, and **c** relevant Pareto fronts, for Bandgap Objective-3, stress sate ratio $b_s = \sqrt{2}$ and aspect ratio 10

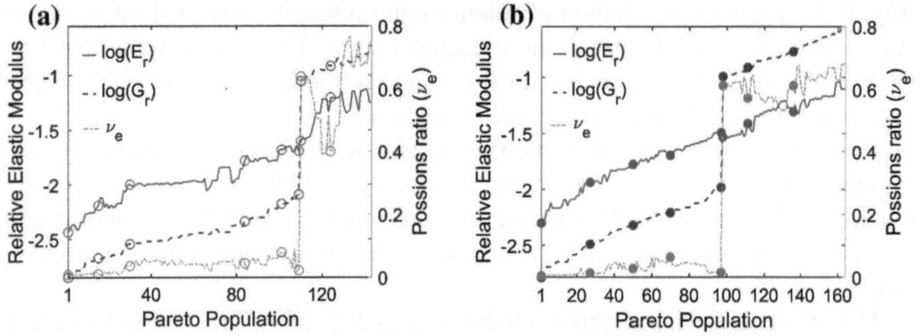

Fig. 6.8 Gradient of relative elastic and shear modulus E_r and G_r and effective Poisson's ratio v_e versus Pareto population for both topology resolutions **a** 32 × 32 and **b** 64 × 64, and relative location of selected topologies (Fig. 6.7) marked by circles

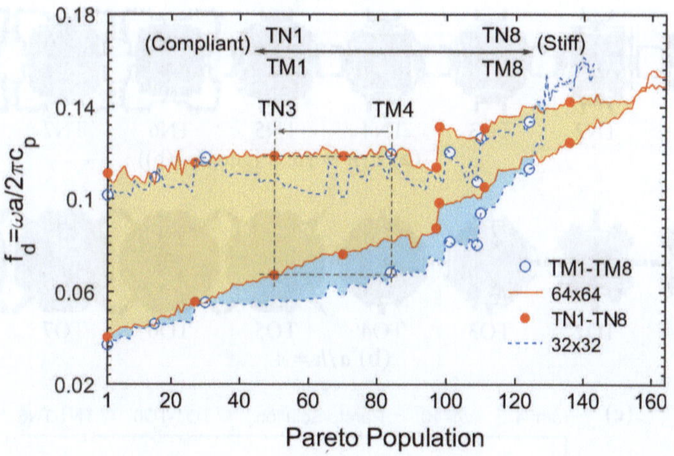

Fig. 6.9 Gradient of bandgap frequency range versus Pareto optimised population of both topology resolutions 32 × 32 and 64 × 64 (Fig. 6.7), and relative location of selected topologies marked by circles

(from left to right) and a sudden jump appears corresponding to the gap in the relevant Pareto fronts. The change in the topology mode leads to much higher effective shear modulus and Poisson's ratio and negligible reduction of effective elastic modulus. However, the significant increase in the shear modulus and Poisson's ratio leads to higher shear and normal stiffness respectively.

The gradient of optimised complete bandgap frequency range across the Pareto fronts is also shown in Fig. 6.9 and bandgap efficiency of relevant coarse and refined topologies are compared. The selected Pareto topologies of Fig. 6.2a, b are also marked to signify their relative bandgap frequency range.

As mentioned earlier, it should be noted that the horizontal axis of Fig. 6.9 is just Pareto optimised topology number and does not show the corresponding coarse and refined topologies. To clarify, the coarse topology TM4 and refined topology TN3 are marked which almost coincide in the relevant Pareto fronts as shown in Fig. 6.7c but are shifted in Fig. 6.9. From Fig. 6.9 it is clear that, as expected, the topology refinement has led to a wider spread of topologies with smoother gradient of frequency bands. Also, the aforementioned change in the topology mode (between coarse topologies TM6 and TM7, and refined topologies TN5 and TN6) provides slightly wider bandgap at a bit higher frequency range (i.e. almost the same RBW), but as shown in Fig. 6.8 with much higher stiffness.

The refined topologies of complete bandgaps of other aspect ratio 4 are further obtained and compared with aspect ratio 10 as shown in Fig. 6.10. The optimised topologies TN1-TN8 of Fig. 6.7b with aspect ratio 10 are shown again in Fig. 6.10a for presentation of their distinction compared with topologies of aspect ratio 4. Like bandgaps of asymmetric mode, the Pareto fronts of complete bandgaps shown in Fig. 6.10c show the considerably higher RBW of thicker unit-cells (of the same

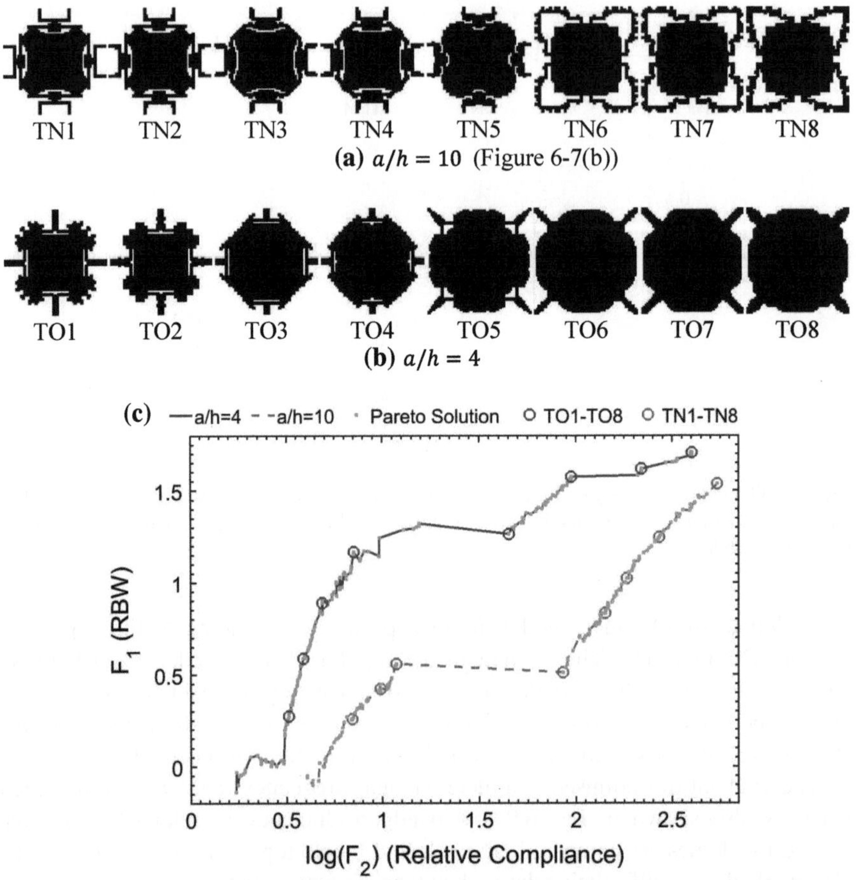

Fig. 6.10 Selected Pareto optimised topologies with resolution 64 × 64 with aspect ratio **a** 10 (repeated from Fig. 6.7b for comparison) and **b** 4, and **c** relevant Pareto fronts for Bandgap Objective-3 and stress sate ratio $b_s = \sqrt{2}$

width) which obviously have higher overall stiffness as well due to their larger thickness. The variation of the topology mode from TO4 to TO6 is evident.

The variation of bandgap frequencies across Pareto front solutions of Fig. 6.10 are also shown in Fig. 6.11. The results clearly show wider bandgaps produced by topologies of aspect ratio 4. In contrast to all other preceding results, the topologies on the stiff side of Pareto front of aspect ratio 4 have bandgap at relatively lower frequency ranges compared with topologies of compliant side.

A bunch of topologies on the extreme stiff side of Pareto front (Right side of Fig. 6.11) are invalid due to having negative bandgap. As stated earlier in this section, this is due to rough modelling of PhP unit-cell during optimisation which overestimates tiny bandgaps for these topologies.

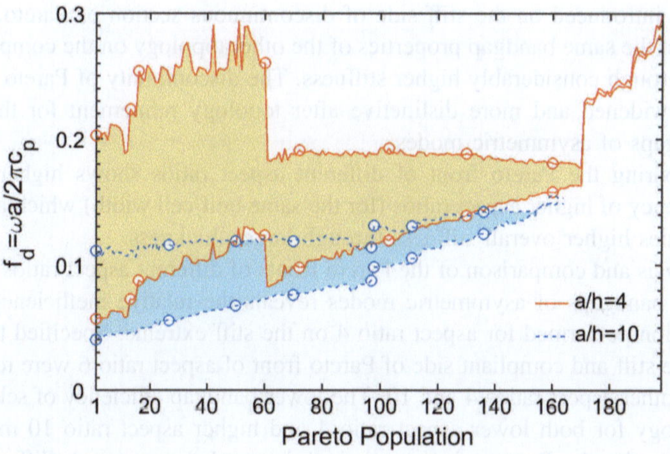

Fig. 6.11 Gradient of bandgap frequency range versus Pareto optimised population for aspect ratios 4, 6 and 10, and relative location of selected topologies marked by circles

6.5 Concluding Remarks

In this chapter, an improved optimisation framework was employed for topology optimisation of PhPs and a topology refinement was performed. The void elements of topology and relevant boundary conditions were excluded from FEM solution stage for faster fitness evaluation. An edge mutation was exclusively performed on the interfacial design variables during the last stage of optimisation to focus on exterior boundaries, and a strict repair operation was applied, in parallel, to avoid checkerboards. Bandgap Objective-1 and Bandgap Objective-3, as introduced in Chap. 5, were investigated for bandgaps of asymmetric modes and complete bandgap of guided wave modes respectively.

Based on the achieved Pareto fronts and analysis of optimised topologies, the following conclusions can be made:

- Due to higher versatility of refined design domain, as expected, the Pareto front of refined topology provides a more uniform distribution of optimised solutions with smoother variation of bandgap-stiffness properties.
- The topology refinement significantly shifted the Pareto front of bandgaps of asymmetric modes towards higher bandgap-stiffness efficiency. However, the Pareto front of complete bandgaps was shifted marginally. This is mainly due to having stricter bandgap objective in the case of complete bandgaps (to open bandgap between all mode types) which restricts the solutions space.
- For extreme topologies on the stiff side of Pareto fronts, complete bandgaps are generally stiffer than bandgaps of asymmetric modes.
- In all cases, a sudden change in the structure of topology leads to a considerable gap i.e. discontinuity in the intermediate section of Pareto front. The topology

mode introduced on the stiff side of discontinuous section of Pareto, provides almost the same bandgap properties of the other topology on the compliant side, but through considerably higher stiffness. The discontinuity of Pareto front was even widened and more distinctive after topology refinement for the case of bandgaps of asymmetric modes.

- Comparing the Pareto front of different aspect ratios shows higher bandgap efficiency of higher aspect ratios (for the same unit-cell width) which apparently provides higher overall stiffness through larger thickness.
- Analysis and comparison of the Pareto fronts of different aspect ratios 10, 6 and 4 for bandgaps of asymmetric modes reveals the relative inefficiency of optimisation performed for aspect ratio 4 on the stiff extreme. Specified topologies on the stiff and compliant side of Pareto front of aspect ratio 6 were remodelled with other aspect ratios 4 and 10. The lower bandgap efficiency of selected stiff topology for both lower aspect ratio 4 and higher aspect ratio 10 may be the reason why the Pareto of aspect ratio 6 has ended up to such different set of topologies. This could be avoided by considering larger population size which of course imposes higher computational cost.
- One may get different topologies by assuming different topology resolution, symmetry and constraints based on the design requirements and/or production limitations. However, in this section of study it is shown that how the topology refinement, while maintaining the solid feature sizes, influences bandgap-structural efficiency.

References

Guest, J. K., Prévost, J., & Belytschko, T. (2004). Achieving minimum length scale in topology optimization using nodal design variables and projection functions. *International Journal for Numerical Methods in Engineering, 61*(2), 238–254.

Guest, J. K., & Smith Genut, L. C. (2010). Reducing dimensionality in topology optimization using adaptive design variable fields. *International Journal for Numerical Methods in Engineering, 81*(8), 1019–1045.

Chapter 7
Optimisation of Porous 2D PhPs for Deformation-Induced Tunability

7.1 Introduction

Due to wide multiscale application of PhCrs it is of great value to introduce controllable phononic bandgaps to be tailored, degraded or enhanced during their function or with switchable bandgap properties (Matar et al. 2013). For instance an adaptive phononic wave guide can be designed so that its functional frequency or direction can be changed during operation. This can be achieved through use of active features, such as tailored changes in topology and/or material properties. For example active piezoelectric material can be used locally to control the effective stiffness of phononic unit-cell interconnections (Bergamini et al. 2014) or spatially gradient piezoelectric polarisation can be introduced to control the wave motion (Rupp et al. 2010). The microstructural composition of phononic unit-cell may also be altered by (for example) rotation of asymmetric scattering inclusions (Goffaux and Vigneron 2001) or the effective mechanical properties may be tuned by orientation of symmetric-anisotropic inclusions (Lin and Huang 2011). Moreover, the thermal sensitivity of mechanical properties and density of constitutive materials is applicable in control of the phononic properties (Yao et al. 2011).

Another interesting approach is to tune the phononic bandgaps through deformation of the phononic lattice under external load, where tunability is obtained through deformation-induced gradient of both topology and stiffness (Gei et al. 2013). The strained geometry and internal stress state of the unit-cell then lead to variation of effective stiffness. The stiffness gradient can be amplified with a base material with nonlinear elastic constitutive properties. Additionally, buckling may be 'designed' into the system at certain external loads to create larger shape changes

Parts of this chapter have been published in the following document:

Hedayatrasa, S., Abhary, K., Uddin, M. S. & Guest, J. K., 2016. Optimal design of tunable phononic bandgap plates under equibiaxial stretch. *Smart Materials and Structures*, *25*(5), p. 055025. https://doi.org/10.1088/0964-1726/25/5/055025.

S. Hedayatrasa, *Design Optimisation and Validation of Phononic Crystal Plates for Manipulation of Elastodynamic Guided Waves*, Springer Theses,
https://doi.org/10.1007/978-3-319-72959-6_7

and thus considerable variation of phononic properties. Traction forces can be mechanically applied on the boundaries via physical contact, or contactless load can be applied for example through remote electromagnetic excitation of ferromagnetic inserts. Since the PhCr is supposed to experience relatively large strains for effective gradient of phononic bandgaps, appropriate constitutive properties with perfect elasticity under large deformation is desirable. Therefore, rubber-like elastomeric polymers (e.g. silicon rubber) with large deformation tolerance and almost instantaneous recoverable elastic strain are ideal candidates for such application.

The mechanics of deformation and post-buckling pattern transformation of periodic elastomeric plates with prescribed arrangement of cylindrical holes, under uniaxial compressive strain of up to 10% were numerically and experimentally investigated by Bertoldi et al. (2008). Further Bertoldi and Boyce (2008) studied instability and bandgap tunability of square array of circular/elliptical voids (or solid inclusions) in an elastomeric background solid under different loading conditions, numerically and analytically. Biaxial and equibiaxial tensile strain (up to 50%) and compressive strain (up to 10%) as well as shear strain were examined. Considering neo-Hookean and Gent hyperelastic models revealed the effect of stress stiffening on bandgap tunability of elastomeric PhCrs. Before instability transformation, higher bandgap gradient was observed for Gent material swith stress stiffening compared to neo-Hookean. However, dramatic post-buckling transformation of neo-Hookean lattice was detected with significant effect on phononic bandgap properties.

Likewise the effect of macromechanical equibiaxial compressive strain of up to 20% on modal band structure of square array of circular holes in an elastomeric background was numerically studied by Wang et al. (2013). Gent hyperelastic model was employed and the effect of saturation constant (indicating the stress stiffening behaviour) on RBW versus strain was studied. The deformation-induced evolution of equal frequency contour i.e. wave front at low frequency levels, was also explored.

A novel tunable phononic bandgap made from magneto-elastic elastomer with circular voids which deforms under electromagnetic body force induced by an external magnetic field was analysed by Bayat and Gordaninejad (2015b). It was shown that the induced magnetic field not only affects the bandgap efficiency but also changes preferential directions of wave propagation.

Wang et al. (2014) introduced a novel switchable locally resonant acoustic metamaterial plate that was trigged by a buckling transformation induced by uniaxial compressive strain of 10%. Cylindrical metallic cores were periodically inserted into elastomeric sleeves and connected to perforated elastomeric background through four perpendicular beams. Rudykh and Boyce (2014) introduced tunable bandgaps trigged by lateral compression of layered media over critical buckling load which leads to wrinkling of interfacial layers and development of phononic bandgaps. Bayat and Gordaninejad (2015a) proposed a 1D phononic crystal bilayer slab produced by a thin film bonded to a thick compliant substrate. Bandgap switchability was observed and achieved due to buckling induced surface wrinkling under a longitudinal compressive strain.

As the topology of PhCr unit-cell predominantly defines the bandgap properties, many research works have examined the use of topology optimisation to maximise RBW. The optimum topology of very thick PhCr has been investigated for bandgaps of bulk waves, and a few recent works have studied the optimum topology of PhCr plates (PhPs) having finite thickness in order to manipulate guided waves. However, very few papers have considered topology optimisation of tunable bandgaps and acoustic metamaterials.

Rupp et al. (2010) worked on polarisation pattern optimisation of a piezoelectric solid plate in order to design switchable wave guides or filters of vibroacoustic waves at specified frequency. Gradient-based optimisation was employed and the distribution of two contrasting material phases (with positive and negative polarisation constants leading to different acoustic impedances) throughout the domain was optimised. Evgrafov et al. (2008) optimised the topology of tunable waveguides with the capability to bend input waves into different ports at a predefined frequency using undeformed and deformed configurations. The spatial distribution of constitutive materials Silicon-Aluminium in a square domain was identified through gradient-based optimisation. An absorbing boundary was considered for all boundaries of the design domain (typically a small part of large elastic body) and unidirectional tensile strain of 0.2% was applied for switching the wave guiding function. To the best of author's knowledge based on the literature review, however, there has been no study on optimising bandgap tunability of PhCrs.

The aim of the present part of the research is to design PhP lattices with maximised bandgap tunability (i.e. gradient) under specified externally imposed deformation while providing initially wide complete bandgap of guided waves, within desired number of modal branches. Wide initial RBW and maximum tunability are typically competing objectives, motivating the use of systematic optimisation. A porous elastomeric PhP unit-cell is considered which is produced by through thickness perforation of a uniform background plate. The relatively low transmission loss of guided waves through thin walled structures makes them an ideal candidate for fabrication of low loss ultrasound devices and structural health monitoring purposes (Olsson Iii and El-Kady 2009; Olsson Iii et al. 2008). Bertoldi and Boyce (2008) studied the phononic bandgap tunability of such lattices under tensile strain but with infinite thickness (plain strain condition) and for specified profile of cylindrical holes. Herein, the topology (i.e. perforation pattern) is optimised to achieve the following simultaneous objectives:

1. Maximum RBW of guided waves in undeformed configuration, and
2. Maximum relative gradient of RBW (with reference to undeformed state) when deformed under macromechanical equibiaxial stretch of up to 10%.

The gradient of bandgap could be explored in an ascending or descending fashion based on the desired application. The first 12 modal branches are preferably considered for evaluation of total RBW which were shown to introduce major complete bandgaps of mixed guided wave modes based on the results presented in Chap. 5. Due to deformation-induced evolution of modal band structure which shifts the bandgap frequencies, a modified relative gradient of phononic bandgap is also evaluated by considering the initial bandgap frequency as a reference value for tunability analysis.

The present study is limited only to phononic tunability produced through nonlinear stable deformation of PhP lattices under assumed tensile stretch. Therefore, lattices exhibiting buckling instabilities are neglected. This provides a stable gradual bandgap tunability via deformation, rather than sudden bandgap switching caused by evolutionary post buckling deformation. Moreover, buckling instabilities lead to considerably large microstructural deformations which may be destructive to a particular constitutive material and application.

A multi-objective genetic algorithm (GA) is employed to search the topological design space, taking into account aforementioned multistate-multimodal objectives. The finite element method (FEM) is used to evaluate the modal band structure and nonlinear deformation of the unit-cells to enable estimation of a fitness function. The Heaviside Projection Method (HPM) (Guest et al. 2004; Guest and Smith Genut 2010) is used to map independent nodal design variable fields onto the unit cell topology in order to control the minimum length scale of designed features, and a dimension reduction scheme is used to adaptively control the design space resolution independently of the FEM mesh resolution (Guest et al. 2004; Guest and Smith Genut 2010).

7.2 Theory and Governing Equations

Square PhP lattices with 2D periodicity comprised of macro-mechanically orthotropic unit-cells with square symmetric topology are considered herein. Plate unit-cells of transversal aspect ratio a/h are modeled where a and h are lattice periodicity and thickness respectively, as shown schematically in Fig. 7.1 for an arbitrary perforation profile. The unit-cell is uniformly perforated through thickness (along z-axis) making planar heterogeneity (in xy-plane), which is potentially able to open phononic bandgaps by constructive scattering of elastodynamic guided waves.

The bandgap width of hypothesized PhP lattice is supposed to be tuned via externally applied equibiaxial tensile stretch of up to 10% along x and y axes. Thus the deformed lattice maintains square symmetry and orthotropic structure in the absence of buckling instability. Thick PhP with aspect ratio $a/h = 2$ is modelled in order to reduce the risk of local lateral buckling under the tuning finite strain. Topology analysis and assessment is performed at the unit-cell scale for the sake of computational cost and so considering appropriate periodic boundary conditions is essential at all phases for reliable calculation of deformed state and modal band structure. The finite element method is employed to model the deformation of the plate's unit-cell under a prescribed strain as well as its modal band analysis in the prestressed state.

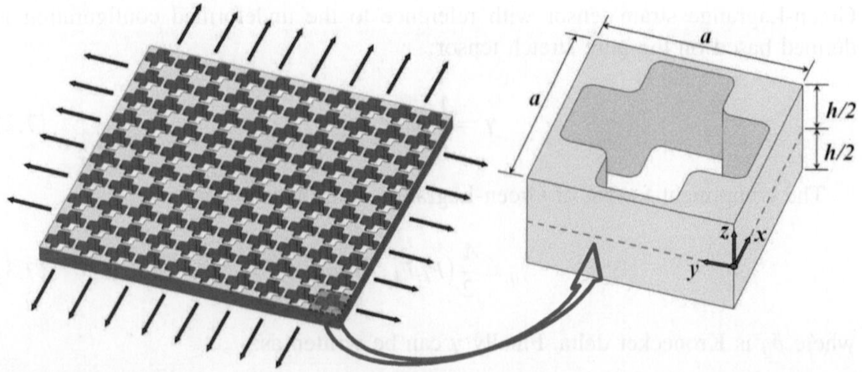

Fig. 7.1 2D PhP lattice and its unit-cell with arbitrary perforation profile to be deformed under uniform equibiaxial tensile stretch; square lattice periodicity a in xy-plane, thickness h along z-axis, and aspect ratio $a/h = 2$

The phononic tunability of PhP under finite strain is desired. So an iterative nonlinear FEM analysis considering a nonlinear hyperplastic material model and geometrical nonlinearities is required. As a result of this initial nonlinear static solution the tangential stiffness of the unit-cell is defined. This tangential stiffness is used for further buckling stability and modal analysis of deformed PhP through a linear perturbation analysis.

ANSYS FEM software (*ANSYS® Academic Research, Release 14.5*) is employed to perform the nonlinear prestress analysis, to check stability, and to perform the modal band analysis of topologies. The governing equations and analysis procedure are detailed in the following sections.

7.2.1 Nonlinear Static Analysis of Prescribed Finite Strain

Since the infinitesimal strain is not able to capture the nonlinear deformation of continuum under large displacements, the Green-Lagrange strain tensor is employed which is defined based on deformation gradient tensor with reference to undeformed state of the medium (de Borst1 et al. 2012).

Assuming solid particles in initial material coordination **X** deformed to current spatial coordination **x**, the deformation gradient tensor would be:

$$\mathbf{F} = \frac{\delta \mathbf{x}}{\delta \mathbf{X}} \tag{7.1}$$

that itself can be decomposed into two pure rotation **R** and pure deformation **U** components such that $\mathbf{F} = \mathbf{R} \cdot \mathbf{U}$. By introducing right Cauchy-Green deformation tensor $\mathbf{C} = \mathbf{F}^T \cdot \mathbf{F} = \mathbf{U}^2$ the rigid body rotation is eliminated and

Green-Lagrange strain tensor with reference to the undeformed configuration is defined based on the pure stretch tensor:

$$\boldsymbol{\gamma} = \frac{1}{2}(\mathbf{C} - \mathbf{I}) \tag{7.2}$$

The component format of Green-Lagrange strain tensor $\boldsymbol{\gamma}$ is:

$$\gamma_{ij} = \frac{1}{2}\left(F_{ki}F_{kj} - \delta_{ij}\right) \tag{7.3}$$

where δ_{ij} is Kronecker delta. Finally $\boldsymbol{\gamma}$ can be written as:

$$\boldsymbol{\gamma} = \mathbf{e} + \boldsymbol{\alpha} \tag{7.4}$$

which $\mathbf{e}$ and $\boldsymbol{\alpha}$ are linear (rotation dependent) and nonlinear (rotation altering) terms of strain $\boldsymbol{\gamma}$ respectively, defined as:

$$e_{ij} = \frac{1}{2}\left(\frac{\partial u_i}{\partial X_j} + \frac{\partial u_j}{\partial X_i}\right) \tag{7.5}$$

$$\alpha_{ij} = \frac{1}{2}\frac{\partial u_k}{\partial X_i}\frac{\partial u_k}{\partial X_j} \tag{7.6}$$

Apparently the definition of linear strain $\mathbf{e}$ [Eq. (7.5)] is exactly like that of infinitesimal strain. So, for small deformation gradients, quadratic term $\boldsymbol{\alpha}$ is negligible and Green-Lagrange strain equals to classic small strain. The strain energy conjugate to $\boldsymbol{\gamma}$ is Second Piola–Kirchhoff (2PK) stress $\mathbf{S}$ and the corresponding true (Cauchy) stress tensor is:

$$\boldsymbol{\sigma} = \frac{1}{det\mathbf{F}}\left(\mathbf{F} \cdot \mathbf{S} \cdot \mathbf{F}^T\right) \tag{7.7}$$

The true strain $\boldsymbol{\varepsilon}$ can also be obtained easily from pure stretch tensor $\mathbf{U}$:

$$\boldsymbol{\varepsilon} = \ln(\mathbf{U}) \tag{7.8}$$

Based on the principle of virtual work the balance of internal and external work over the undeformed volume V_0 induced by virtual displacement increment $\delta\mathbf{u}$ from the undeformed configuration and relevant strain tensor $\delta\boldsymbol{\gamma}$ for static deformation under traction force $\mathbf{t}$ on the undeformed surface S_0 is:

$$\int_{V_0} \delta\boldsymbol{\gamma}^T\mathbf{S}dV = \int_{S_0} \delta\mathbf{u}^T\mathbf{t}dS \tag{7.9}$$

where $\boldsymbol{\gamma} = \boldsymbol{\gamma}_L + \Delta\boldsymbol{\gamma}$ and $\mathbf{S} = \mathbf{S}_L + \Delta\mathbf{S}$ indicate the increment of stress and strain from the last known configuration at the beginning of a load-step (indicated by subscript L). For small increments a linear relation between $\Delta\boldsymbol{\gamma}$ and $\Delta\mathbf{S}$ can be assumed by instantaneous stiffness of material $\mathbf{D} = \Delta\mathbf{S}/\Delta\boldsymbol{\gamma}$. Then by definition of displacement field $\mathbf{u}$ based on vector of nodal displacement $\mathbf{q}$ and matrix of shape functions $\mathbf{N}$:

$$\mathbf{u} = \mathbf{N} \cdot \mathbf{q} \tag{7.10}$$

and by decomposition of Green-Lagrange strain increment $\Delta\boldsymbol{\gamma} = \Delta\mathbf{e} + \Delta\boldsymbol{\alpha}$, linearised FEM equation will be obtained:

$$\mathbf{K}_t^{n,i}\Delta\mathbf{q}^{n,i+1} + \mathbf{f}_{int}^{n,i} = \mathbf{f}_{ext}^{n} \tag{7.11}$$

where for iterative analysis of nonlinear static deformation the goal is to find incremental nodal displacement $\Delta\mathbf{q}$. The first superscript shows loadstep number and the second one stands for sub-step iteration number. $\mathbf{K}_t$ is the tangential stiffness matrix of FEM model:

$$\mathbf{K}_t = \mathbf{K} + \mathbf{K}_s \tag{7.12}$$

where $\mathbf{K}$ is the linear stiffness matrix related to linear displacement based on current geometry of model:

$$\mathbf{K} = \int_V \mathbf{B}^T\mathbf{D}\mathbf{B}dV \tag{7.13}$$

and $\mathbf{K}_s$ is the stress stiffness matrix related to the contribution of nonlinear geometric displacement to the tangential stiffness and indicating the resistance of internal forces to deformation:

$$\mathbf{K}_s = \int_V \mathbf{B}_s\boldsymbol{\tau}\,\mathbf{B}_s dV \tag{7.14}$$

where $\mathbf{B}$ is the linear strain-displacement matrix such that $\Delta\mathbf{e} = \mathbf{B} \cdot \Delta\mathbf{q}$, $\mathbf{B}_s$ is a matrix of shape function derivatives and $\boldsymbol{\tau}$ is a matrix of true (Cauchy) stresses (Filipovic et al. 2008; de Borst1 et al. 2012). The internal force $\mathbf{f}_{int}$ is also formulated as:

$$\mathbf{f}_{int} = \int_V \mathbf{B}^T\boldsymbol{\sigma}dV \tag{7.15}$$

The external load is applied in predefined load-steps and the convergence of solution at each load-step is achieved by sub-step increments until reaching acceptable difference between load-step external loads and internal (restoring) forces. The tangential stiffness matrix and internal force matrix are updated at the end of each sub-step iteration and new iteration is performed by applying residual load $\left(f_{ext}^{n} - f_{int}^{n,i}\right)$ when convergence is not achieved for current loadstep.

In this study the PhP unit-cell is externally loaded by a macromechanical displacement load through intended equibiaxial tensile deformation $\mathbf{F}_d$ on the unit-cell boundaries:

$$\mathbf{F}_d = \begin{bmatrix} \xi_d & 0 & 0 \\ 0 & \xi_d & 0 \\ 0 & 0 & \xi_d \end{bmatrix} \quad (1 < \xi_d < 1.1) \tag{7.16}$$

so that the unit-cell size a after deformation would be $a_d = \xi_d a$. In order to satisfy the periodicity of unit-cell under assumed macromechanical deformation, a periodic displacement load is applied on corresponding nodes of opposing boundaries such that:

$$\mathbf{u}(\mathbf{X} + \mathbf{A}_b) = \mathbf{u}(\mathbf{X}) + \mathbf{F}_d\mathbf{A}_b \tag{7.17}$$

where $\mathbf{u} = \{u\ v\ w\}^T$ is the displacement vector and $\mathbf{A}_b$ is the lattice spanning vector of unit-cell opposing boundaries:

$$\mathbf{A}_b = \{\mathrm{a}\ 0\ 0\}^T \quad \text{for boundaries parallel to y-axis, or} \tag{7.18}$$

$$\mathbf{A}_b = \{0\ a\ 0\}^T \quad \text{for boundaries parallel to x-axis} \tag{7.19}$$

To solve this displacement control analysis defined by constraint equations (7.18) and (7.19), the problem is reformulated through Lagrangian multiplier method:

$$\begin{bmatrix} \mathbf{K}_t & \mathbf{L} \\ \mathbf{L}^T & 0 \end{bmatrix} \begin{bmatrix} \Delta\mathbf{q} \\ \Delta\boldsymbol{\psi} \end{bmatrix} = \begin{bmatrix} -\mathbf{f}_{int} \\ \Delta\mathbf{c} \end{bmatrix} \tag{7.20}$$

which is the matrix format of the following linearised equations:

$$\mathbf{K}_t^{n,j}\Delta\mathbf{q}^{n,j+1} = -\mathbf{L}\Delta\psi^{n,j+1} - \mathbf{f}_{int}^{n,j} \tag{7.21}$$

$$\mathbf{L}^T\Delta\mathbf{q}^{n,j+1} = \Delta\mathbf{c}^{n,j+1} \tag{7.22}$$

Equation (7.21) is actually the balance of momentum under equivalent external load:

$$\mathbf{f}_{ext}^{n,j+1} = -\mathbf{L}\Delta\boldsymbol{\psi}^{n,j+1} \tag{7.23}$$

corresponding to the periodic boundary condition defined by Eq. (7.22) applying incremental periodic displacement $\Delta\mathbf{c}$ on constrained degrees of freedom by matrix

L. So the external load is implicitly defined and updated in all sub-steps. The displacement increment of each load-step $\Delta\mathbf{c}^n$ is totally applied at the first sub-step and no further displacement is applied during the next sub-steps until a desired residual load tolerance is achieved:

$$\Delta\mathbf{c}^{n,j} = \begin{cases} \Delta\mathbf{c}^{n,j} & (j = 1) \\ 0 & (j > 1) \end{cases} \tag{7.24}$$

7.2.2 *Hyperelastic Constitutive Model*

The constitutive material properties play an essential role in maintaining the recoverability of the tunable bandgaps under finite macromechanical strain. Therefore, silicon rubber is selected as the background material of tunable PhP as it can tolerate large deformation with almost instantaneous recoverable elastic strain.

Hyperelasticity is commonly used as constitutive model of rubber like materials by requiring the strain energy density W to be a scalar function of strain tensor γ. For an isotropic material the strain energy density becomes a function of stretch invariants I_1, I_2 and I_3 only:

$$W = W(I_1, I_2, I_3) \tag{7.25}$$

where the invariants are functions of eigenvalues λ_1, λ_2 and λ_3 of right Cauchy-Green deformation tensor $\mathbf{C}$:

$$I_1 = \lambda_1^2 + \lambda_2^2 + \lambda_3^2, \quad I_2 = \lambda_1^2\lambda_2^2 + \lambda_2^2\lambda_3^2 + \lambda_3^2\lambda_1^2, \quad I_3 = J^2 = \lambda_1^2\lambda_2^2\lambda_3^2 \tag{7.26}$$

The Gent Hyperelastic model is employed herein which defines strain energy density as:

$$W = -\frac{\mu J_m}{2}\ln\left(1 - \frac{\bar{I}_1 - 3}{J_m}\right) + \frac{k_b}{2}\left(\frac{J^2 - 1}{2} - \ln J\right) \tag{7.27}$$

μ is initial shear modulus, k_b is initial bulk modulus and J_m is a limiting value concerned with strain saturation of hyperelastic material. $\bar{I}_1$ is modified invariant pertaining to volume preservative part of right Cauchy-Green deformation tensor $\widetilde{\mathbf{C}}$:

$$\widetilde{\mathbf{C}} = I_3^{-1/3}\mathbf{C} \tag{7.28}$$

such that:

$$\widetilde{I}_p = I_3^{-p/3} I_p \quad (p = 1, 2, 3) \tag{7.29}$$

The 2PK stress tensor $\mathbf{S}$ as the strain energy conjugate of Green-Lagrange strain tensor $\boldsymbol{\gamma}$ is:

$$\mathbf{S} = \frac{\partial W}{\partial \boldsymbol{\gamma}} = 2\frac{\partial W}{\partial \mathbf{C}} \tag{7.30}$$

by which the true Cauchy stress tensor $\boldsymbol{\sigma}$ [Eq. (7.7)] and internal force $\mathbf{f}_{int}$ [Eq. (7.15)] can be defined. Then the tangential material property matrix needed to calculate stiffness matrix $\mathbf{K}$ [Eq. (7.13)] during nonlinear solution equilibrium iterations, is defined by:

$$\mathbf{D} = \frac{\partial \mathbf{S}}{\partial \boldsymbol{\gamma}} = 4\frac{\partial^2 W}{\partial \mathbf{C}^2} \tag{7.31}$$

7.2.3 Buckling Stability Analysis

An important phase in assessing randomly generated PhP unit-cells during optimisation algorithm is to evaluate the buckling stability of topology under desired finite strain. Despite assumed macromechanical biaxial tensile displacement load which is generally unlikely to cause structural instabilities, this evaluation is vital due to uncertainty in topology and loading conditions of microstructural segments of random unit-cells to be examined. Buckling of unit-cell leads to evolutionary change of topology and its modal band structure. In this case a post-buckling deformation analysis is necessary by distorting the unit-cell based on relevant buckling mode shape and performing further nonlinear deformation analysis. However, the aim of this portion of study is to achieve tunability through stable deformation of phononic lattice.

As explained earlier in Sect. 7.2.1, the intended equibiaxial stretch is applied to the unit-cell through periodic boundary conditions which, unlike uniform boundary conditions, provides flexibility for asymmetric deformation of unit-cell. So, the earliest indication of buckling is un-convergence of nonlinear static solution because of instability initiated by likely shape distortion under tuning stretch.

In case of successful nonlinear deformation analysis, a linear perturbation Eigenvalue buckling analysis should be performed to make sure about reliable stable deformation of unit-cell under final tuning stretch. To serve this purpose, a small periodic perturbation stretch is additionally applied to the deformed unit-cell and relevant bucking loads are defined based on the tangential stiffness matrix of FEM model $\mathbf{K}_t$ in deformed state.

A small fraction $(0 < c_b < 1)$ of the final external load $\mathbf{f}_\mathrm{d}$ applied in nonlinear static analysis (to achieve desired finite strain) is assumed as the external load required for intended small perturbation stretch and the relevant nodal displacement tensor is defined by linear solution:

$$\mathbf{K}_{\mathrm{t}}\mathbf{q}_{\mathrm{p}} = c_b\mathbf{f}_{\mathrm{d}} \quad (0 < c_b < 1) \tag{7.32}$$

So perturbation induced stress tensor is calculated and relevant stress stiffness matrix $\mathbf{K}_{\mathrm{p}}$ is calculated from Eq. (7.14). Finally the eigenvalue analysis is performed for non-trivial solutions of proportional equilibrium between internal resisting forces and stress stiffening loads:

$$\mathbf{K}_{\mathrm{t}}\mathbf{q} = \eta\mathbf{K}_{\mathrm{p}}\mathbf{q} \tag{7.33}$$

A set of eigenvalues η_j and relevant eigenvectors (mode shapes) $\mathbf{q}_j$ are calculated. The Eigenvalue η_j is actually the load factor defining critical buckling load $\mathbf{f}_{cr}$ of *j*th buckling mode:

$$(\mathbf{f}_{cr})_j = (1 + \eta_j c_b)\mathbf{f}_d \tag{7.34}$$

Principally only the first buckling mode is checked as the higher order modes are unlikely to happen. So the model is stable provided that the critical load of first mode is greater than applied external load or is of opposite direction (i.e. is negative):

$$(\mathbf{f}_{cr})_1 > (1 + c_b)\mathbf{f}_d \quad \text{or} \quad (\mathbf{f}_{cr})_1 < 0 \tag{7.35}$$

In fact, the small perturbation load factor c_b embedded in stability condition ensures existence of a stable strain range above the desired tuning finite strain.

7.2.4 *Modal Band Analysis*

The key part of the analysis is to determine the modal band structure of the PhP unit-cells, which serves as the objective function for the optimisation problem. For this purpose Bloch-Floquet periodic boundary conditions are applied and the modal frequencies of unit-cell at different wave vectors are calculated following the common procedure already discussed in Chap. 3. The linear stiffness matrix of the undeformed unit-cell and the tangential stiffness matrix of deformed state are used independently for calculation of relevant modal responses.

7.3 Topology Optimisation Methodology

A multi-objective topology optimisation algorithm is used together with external calls to the FEM software ANSYS to evaluate the candidate topology.

The two objective functions of the problem are first discussed, followed by a discussion of the optimisation algorithm and details regarding efficient FEM modelling of PhP.

7.3.1 Objective Functions

Two objectives are considered herein: widest acoustic bandgap in initial (undeformed) state and maximum tunability of PhP within specified number of modal branches. As mentioned earlier, the goal is to design specialised phononic structures so that despite having wide bandgaps initially, show significant bandgap gradient when subjected to specified finite tensile strain. The gradient of bandgap could be explored in an ascending or descending fashion based on the desired application.

The definition of first objective concerning maximum bandgap width is straightforward. For the undeformed state of PhP, the widest bandgap at lowest frequency range is desired such that the largest controllable wavelength through specified unit-cell size is achieved. The first objective function F_1 is therefore defined as (Sigmund and Jensen 2003):

$$F_1 = \sum_{j=2}^{n_m} \Omega_E(j) = \sum_{j=2}^{n_m} \frac{\min_{i=1}^{n_k} \omega_j^2(k_i) - \max_{i=1}^{n_k} \omega_{j-1}^2(k_i)}{0.5\left(\min_{i=1}^{n_k} \omega_j^2(k_i) + \max_{i=1}^{n_k} \omega_{j-1}^2(k_i)\right)} \tag{7.36}$$

where F_1 is cumulative eigenvalue RBW for subsequent modal branches $(j-1)$ and (j) up to the upper mode of interest $j = n_m$ over $i = n_k$ discrete search points k_i on the border of irreducible Brillouin zone.

As for the second objective, the maximum relative difference of F_1 pertaining to undeformed state ($\xi = 1$) and deformed state ($\xi = \xi_d$) is desired with respect to the undeformed one. Hence the second objective F_2 is formulated as:

$$F_2 = \frac{\left(F_1|_{\xi=1}\right)^2 - \left(F_1|_{\xi=\xi_d}\right)^2}{\left(F_1|_{\xi=1}\right)^2} \tag{7.37}$$

This quadratic definition of the relative gradient proved to provide a better distribution of optimised topologies across the Pareto front. It should be noted that F_2 could be maximised or minimised to achieve topologies with enhanced descending or ascending bandgap gradient respectively.

Applying tensile deformation on the unit-cell shifts the modal frequencies (generally to higher magnitudes) and the relevant bandgaps do not necessarily correspond to the same frequency range as the undeformed state. So it is appropriate to consider the initial bandgap frequency of unit-cell as a reference to define bandgap gradients. Furthermore, the deformation-induced evolution of the modal band structure of random topologies may lead to development of new bandgaps between higher modes compressed at the same frequency bandgap as the undeformed state.

Consequently, a modified form of second objective function F_2 [Eq. (7.37)] can be defined to manage these conditions by improved calculation of F_1 in the deformed state. To serve this purpose: (i) the maximum bandgap frequency of undeformed unit-cell ω_R within modal branches of interest n_m is taken to introduce an initial reference frequency range 0 to ω_R, and (ii) additional number of modal branches n_a is included in calculation of F_1 in the undeformed state to account for the cases where new bandgaps are pressed into this reference frequency range. Updated form of the second objective function F_2 then becomes:

$$\widetilde{F}_2 = \frac{\left(F_1|_{\xi=1}\right)^2 - \left(\widetilde{F}_1|_{\xi=\xi_d}\right)^2}{\left(F_1|_{\xi=1}\right)^2} \tag{7.38}$$

where $\widetilde{F}_1$ is:

$$\widetilde{F}_1 = \sum_{j=2}^{n_m+n_a} \frac{\min_{i=1}^{n_k} \tilde{\omega}_j^2(k_i) - \max_{i=1}^{n_k} \omega_{j-1}^2(k_i)}{0.5\left(\min_{i=1}^{n_k} \tilde{\omega}_j^2(k_i) + \max_{i=1}^{n_k} \omega_{j-1}^2(k_i)\right)} \tag{7.39}$$

$$\tilde{\omega}_j(k_i) = \begin{cases} \omega_j(k_i), & \omega_j(k_i) < \omega_R \\ \omega_R, & \omega_j(k_i) \geq \omega_R \end{cases} \tag{7.40}$$

7.3.2 Multi-objective Optimisation Strategy and Algorithm

A non-dominated sorting genetic algorithm (NSGA-II) is used in this research to explore the design space of tunable PhP unit-cells. The algorithm works directly on a binary representation of topology with distinct material properties to be mapped to the discretised design domain. Genetic algorithms are initiated by a stochastic population of potential solutions, reproduce improved generations (on average) through crossover and mutation functions and terminate once improved solutions are no longer found (or after a certain number of iterations) (Chapman et al. 1994; Hajela et al. 1993; Weise 2006). Non-dominated sorting genetic algorithm (NSGA-II) is adopted to handle the two simultaneous objectives of problem (Pratap et al. 2002). NSGA-II sorts the GA population based on non-dominated fitness and crowding distance. Therefore designs with the most dominated fitness are placed at the first rank (Pareto front); and those with highest crowding distance are preferred to ensure even distribution of solutions within Pareto front (Pratap et al. 2002).

GA has proved competency in topology optimisation of PhCrs and numerous works have reliably employed it in this area (Bilal and Hussein 2012; Dong et al.

2014a, b; Gazonas et al. 2006; Hussein et al. 2006, 2007; Liu et al. 2014; Manktelow et al. 2013). Dong et al. (2014) used multi-objective GA for design of 2D PhCrs with maximum RBW of bulk waves and minimum mass.

Following the procedure already developed in Chap. 5, robust topology initialisation, filtration and modifications are applied to make optimised porous topologies converge towards feasible connected ones without disturbing the randomness and diversity of GA search domain. Accordingly, the adopted NSGA-II algorithm covering these techniques is briefly described as follows:

1. Create a random bit-string initial population for the porous (solid-void) problem mapping topology to FEM model. An identical fully solid population is randomly mutated with specified probability 0.2 to deliver an initial population of likely connected topologies.
2. Pass individuals for fitness evaluation provided that the relevant topology *includes* valid segment touching boundary and with face to face connected elements. Else, degrade the topology by assigning very low fitness for both objectives. The objective functions are to be determined for passed individuals by considering the valid boundary connected segment *only*.
3. Calculate the bandgap objective F_1. If $F_1 \leq 0$ (i.e. no bandgap) then the topology is degraded by low fitness, else it is passed for evaluating second objective (in this way the Pareto front merely delivers the spread of topologies with initial bandgap characteristic in undeformed state).
4. Calculate the tunability objective F_2 (or modified one $\widetilde{F}_2$) provided that the passed topology shows no buckling instability under specified tuning deformation. Else it is degraded by low fitness.
5. Sort the population based on the rank and crowding distance of individuals (Pratap et al. 2002).
6. Create parent pool through tournament selection. Two randomly selected individuals are compared and the one with higher rank is selected as candidate parent. If both have the same rank, the one with highest crowding distance is selected.
7. Create offspring population through single point crossover with certain probability. Two Children are produced per randomly selected parents.
8. Mutate gens of offspring population with specified probability.
9. Modify topology for checkerboard elimination. Randomly selected pixel is inversed (i.e. $0 \rightarrow 1$ and vice versa) if surrounded by more than 2 dissimilar adjacent face to face pixels out of 4 (this well-known technique alleviates the checker board problem which leads to overestimated stiffness when linear elements are used).
10. Modify topology in order to eliminate invalid disconnected segments of topology (Randomly selected disconnected segment, including any one with hinge connection, is eliminated with specified probability).
11. Produce intermediate population by accumulating offspring and current population and sort them based on ranking and crowding distance.

12. Select individuals for the next generation first based on their ranking and second, if the first rank is bigger than required population, based on their crowding distance.
13. Go to Step-2 if termination condition (prescribed number of generations) is not satisfied, otherwise stop.

The contribution and advantages of different steps of implemented algorithm in handling, morphology assessment and evolution of porous topologies are adequately demonstrated in Chap. 5.

7.3.3 FEM and GA Settings

Efficient FEM modelling of a unit-cell is a crucial phase of optimisation procedure, and GA settings also greatly affect the performance of optimisation algorithm and optimality of results. The hypothesized square symmetric PhP unit-cell is modelled by a uniform mesh of square cylindrical elements (ANSYS linear solid element SOLID185) with in-plane discretisation of 40 × 40 finite elements (Fig. 7.2a).

For a relatively thick PhP of interest with assumed transversal aspect ratio of 2, an appropriate transversal mesh size (through the thickness) should be considered for adequate accuracy of FEM analysis. However according to the results presented in Chap. 5 considering only two element layers along the thickness of such plate unit-cell (for the sake of computational intensity) provides acceptable accuracy for estimation of modal band structure of first few modes for fitness evaluation purpose. Following the optimisation, the bandgap efficiency and tunability performance of optimised PhP topologies are evaluated by FEM models with planar resolution of 80 × 80, and 20 element layers through the thickness, to confirm the predicted performance of topologies does not change with mesh refinement.

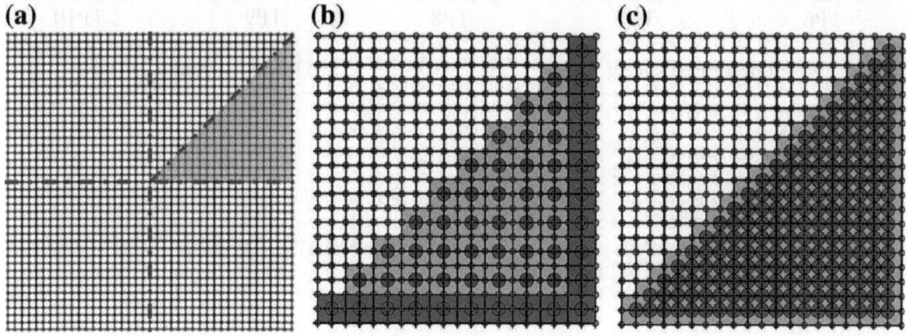

Fig. 7.2 **a** Uniform FEM mesh of square unit-cell with in-plane resolution of 40 × 40, **b** nodal design variable space considered for reduced square symmetric region and **c** corresponding refined design variable space; small circles display FEM nodes and big circles display design variables

Another noteworthy issue concerning FEM modelling of porous PhP is how to model void elements i.e. perforated areas of the unit-cell (vacuumed or air filled). Due to the discrete nature of the GA, boundary conditions can be adapted for each candidate topology, and thus void elements need not be included in the FEM. This has the benefits of avoiding spurious behaviour in void elements and reducing computational expense of the FEM due to a reduced number of degrees-of-freedom.

As square symmetric solutions are desired, only one-eighth section of the PhP unit cell as highlighted in Fig. 7.3a is considered as the independent design domain. To further reduce the number of independent design variables, which stochastic search algorithms are particularly sensitive too, the reduced dimension approach of Guest and Smith Genut (2010) is used. The approach is based on the projection method which controls the minimum length scale of features. The constitutive material is projected from a design variable onto finite element centroids in a radial manner over the prescribed length scale (Guest et al. 2004).

In this study nodal design variables are defined with a small projection radius of one element size so that any design variable fills its four surrounding elements. A low resolution design variable field of size 55, with design variables on every other node, is initially introduced as shown in Fig. 7.2b. After substantial

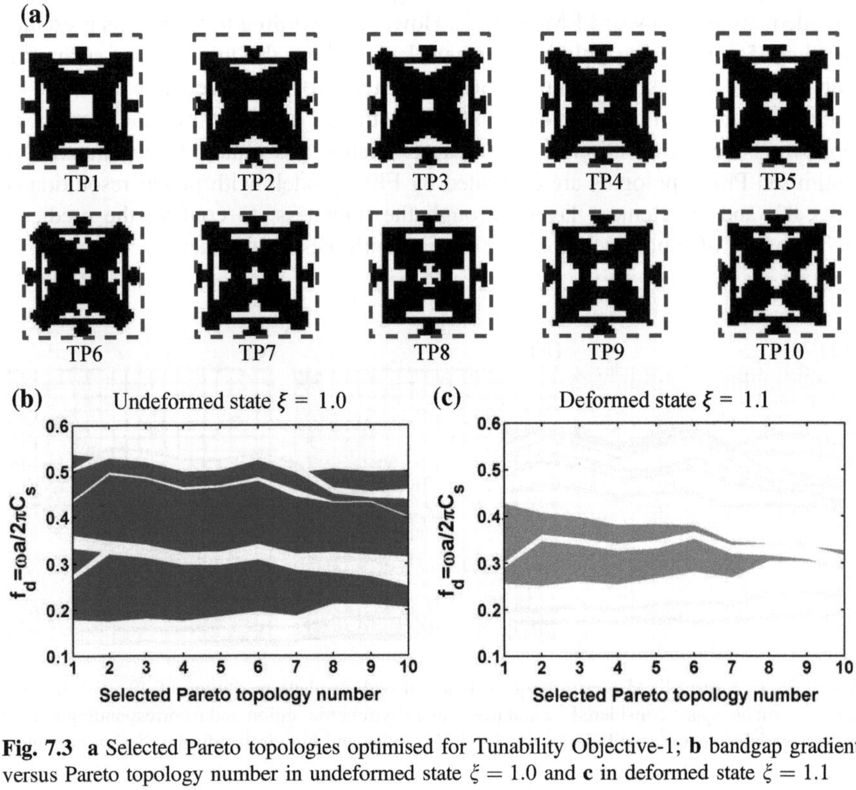

Fig. 7.3 **a** Selected Pareto topologies optimised for Tunability Objective-1; **b** bandgap gradient versus Pareto topology number in undeformed state $\xi = 1.0$ and **c** in deformed state $\xi = 1.1$

Table 7.1 GA settings used for topology optimisation of a tunable PhP

Stage	Population size	Design space size	Probability				Generations
			Crossover	Mutation	Edge mutation	Segment elimination	
1	100	55	0.9	0.3–0.02	–	0–1	135
2	100	190	0.5	0.01–0	0.1–0.01	0.5–1	50

optimisation of topology with this reduced design variable space, design variables are added to the domain (without immediate effect to the topology) creating refined design variable space of size 190 shown in Fig. 7.2c. The boundary nodes are preferably excluded from design variable field in order to limit the radius of orthogonal interconnecting elements of topology (darker elements in Fig. 7.2b) to higher magnitude of two element sizes.

Specific GA operation settings are applied in a dynamic multistage manner, as given in Table 7.1, for efficient optimisation of porous PhP through aforementioned NSGA-II algorithm. According to Table 7.1, the topology optimisation is mainly performed in Stage-1 for reduced design variable space of 135 generations followed by refinement optimisation Stage-2 of 50 generations. GA population size of 100 is chosen for all stages of optimisation. In Stage-1 the design space is initially searched by high mutation probability 0.3 for only five generations to possibly have a good resource of initiating population. Then mutation probability is progressively decreased for finer search during optimisation. Segment elimination probability is added and gradually increased to avoid disturbance of genomes in the early stages of optimisation.

When transferring to refinement Stage-2, the design space size significantly increases from 55 independent design variables to 190. However the interior regions of the solid and void area are filled with many redundant design variables with no effect on the topology. This is due to overlap of projected areas by adjacent design variables in refined design space (Fig. 7.2c) (Guest and Smith Genut 2010). However no major change in topology is expected and the objective is to just refine the boundaries of Pareto topologies of Stage-1. Consequently, the population size is maintained the same, crossover probability is reduced to 0.5 and mutation is predominantly performed on the design variables pertaining to the edge of topology. The probability of this particular mutation (*Edge Mutation* in Table 7.1) is varied from 0.1 to 0.01 throughout Stage-2 for efficient convergence to an optimised Pareto front.

7.4 Optimised Tunable Bandgaps

The optimised Pareto topologies obtained through aforementioned multi-objective optimisation procedure are presented and discussed in this section. Equibiaxial tensile stretch $\xi = 1.1$ is applied as tuning deformation, and perturbation load factor

$c_b = 0.1$ (Sect. 2.3) is considered for buckling stability evaluation purpose. Hyperelastic silicon rubber (Bertoldi and Boyce 2008) with initial shear modulus $\mu = 1.08 \times 10^6$ N/m^2, Poisson's ratio $\nu = 0.4997$ (bulk modulus $k = 2.0 \times 10^9$ N/m^2) and density $\rho = 1050$ kg/m^3 is considered as the base material of PhP and relatively low stress stiffening is assumed by defining $J_m = 0.5$ for Gent hyperelastic model. Lower stress stiffening causes lower material non-linearity and so naturally mitigates deformation-induced bandgap gradient, as confirmed by Bertoldi and Boyce (2008). PhP topology which is optimised based on this particular form of constitutive material, should exhibit higher deformation-induced bandgap gradient in higher stress stiffening condition (i.e. material with lower values of J_m) and of course larger tuning load. In order to study this matter, selected topologies optimised based on $J_m = 0.5$ are remodeled by $J_m = 0.1$ and the effect of material nonlinearity on bandgap tunability efficiency and tuning load is presented.

The first objective of optimisation (i.e. initial bandgap objective) as introduced in Sect. 3.1 is to maximise total RBW F_1 [Eq. (7.36)] within desired number of modal branches n_m. With regard to the results of Chap. 5, F_1 is preferably calculated over the first 12 modal branches ($n_m = 12$) which are shown to introduce major complete bandgaps of mixed guided wave modes. Following Sect. 3.1, the second objective (i.e. Tunability Objective) is defined with four different options as follows:

1. Tunability Objective-1: Maximising deformation-induced RBW gradient F_2 [Eq. (7.37)], so that initially maximised RBW is degraded by deformation and is ideally closed.
2. Tunability Objective-2: Maximising modified deformation-induced RBW gradient $\widetilde{F}_2$ [Eq. (7.38)] by considering additional modal branches ($n_a = 5$), so that initially maximised RBW is degraded by deformation and is ideally closed (with reference to bandgap frequency range of undeformed state).
3. Tunability Objective-3: Minimising deformation-induced RBW gradient F_2 [Eq. (7.37)] so that initially maximised RBW is degraded by deformation and is ideally closed.
4. Tunability Objective-4: Minimising modified deformation-induced RBW gradient $\widetilde{F}_2$ [Eq. (7.38)], by considering 5 additional modal branches ($n_a = 5$), so that initially maximised RBW has minimised sensitivity to the deformation (with reference to bandgap frequency range of the undeformed state) or is even widened through deformation.

In each case a set of 10 Pareto topologies are selected and evaluated so that a variety of basic optimised topology modes between the two Pareto extremes is captured. These topology sets are named TA to TD for the 3 aforementioned tunability objectives. One Pareto extreme corresponds to topology with widest initial RBW, and the other Pareto extreme corresponds to topology with maximum or minimum deformation-induced RBW gradient. Dimensionless modal band structure of PhP unit-cell is defined by dimensionless frequency $f_d = \omega a / 2\pi C_s$, where $C_s = \sqrt{\mu/\rho} = 32.07$ m/s is the shear wave speed in pure solid material in undeformed state.

7.4.1 Maximised RBW Gradient

The results concerning maximised Tunability Objective-1 and its modified Tunability Objectives-2 are presented in this section. The selected Pareto topologies TP1 to TP10 pertaining to Tunability Objective-1 are displayed in Fig. 7.3a. The gradient of relevant complete bandgaps opened within first 12 plate modes of interest are then defined as shown in Fig. 7.3b, c, for undeformed and deformed states respectively. Moreover, the total RBW and relevant tuning macroscopic stress of extreme topologies TA1 and TA10 versus applied macroscopic stretch are presented in Fig. 7.4.

The modal band structures of these extreme topologies in both undeformed and deformed states are also shown in Fig. 7.5. Symmetric and asymmetric plate modes are highlighted by solid and dash lines respectively, and complete bandgap frequency range is highlighted by shaded area. According to Fig. 7.3b, in the undeformed state, five complete bandgaps of extreme topology TP1 in the range of $f_d = 0.18-0.54$ gradually reduce to 4 bandgaps in the range of $f_d = 0.21-0.50$ for other extreme topology TP10, so that total frequency RBW of $\Omega_F = 0.96$ drops to 0.61 across Pareto front (Fig. 7.4). In the deformed state as per Fig. 7.3c, two bandgaps of TP1 in the range of $f_d = 0.26-0.43$ gradually decrease to a single small bandgap of TP10 around $f_d = 0.33$. In this deformed state total frequency RBW of TP1 is 0.48 while that of TP10 is quite completely degraded to 0.03 (Fig. 7.4).

Consequently, as a result of prescribed deformation, the RBW of extreme topology TP1 which has the widest undeformed bandgap, reduces from 0.96 by half to 0.48 (Fig. 7.4a), and RBW of another extreme topology TP10 with around 30%

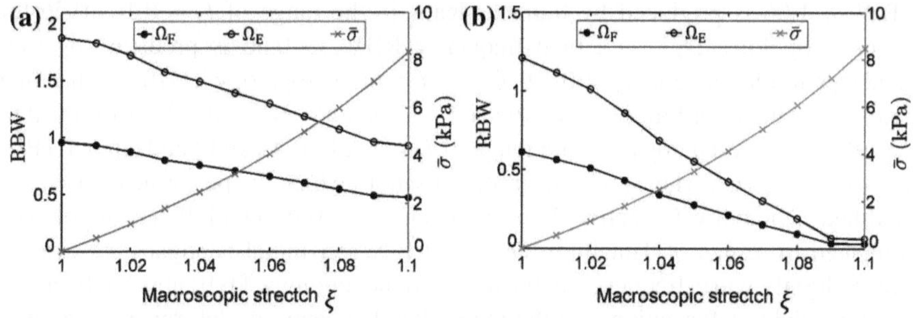

Fig. 7.4 Gradient of total RBW of first 12 plate modes and average macroscopic tuning stress $\bar{\sigma}$, versus macroscopic stretch of up to $\xi = 1.1$ or extreme Pareto topologies **a** TP1 and **b** TP10 (Fig. 7.3a)

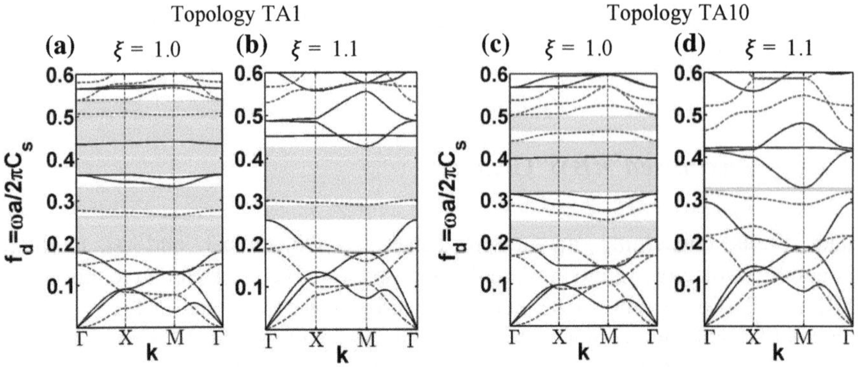

Fig. 7.5 Modal band structure and complete bandgaps of extreme Pareto topologies **a**, **b** TP1 and **c**, **d** TP10 (Fig. 7.4a) in undeformed state $\xi = 1.0$ and deformed state = 1.1; Solid and dash lines stand for symmetric and asymmetric plate modes respectively, and shaded area highlights complete bandgaps

lower initial bandgap of 0.61 is almost closed (Fig. 7.4b). According to the relevant modal band structures depicted in Fig. 7.5, as expected, modal branches are generally shifted up by deformation due to initial tensile stress state of the deformed unit-cell. However, the modal bandgaps are predominantly tuned i.e. degraded by expansion of symmetric plate modes mainly in second and third Brillouin zone borders XM and MΓ.

Likewise, the results concerning maximised Tunability Objective-2 (topologies TQ1 to TQ10) are presented in Figs. 7.6, 7.7, 7.8, and 7.9. As discussed earlier, for this modified objective, RBW of the deformed state is calculated with reference to maximum bandgap frequency of the undeformed state such that the main target is set to degrade the bandgap within its initial frequency range. Moreover, five additional modal branches are calculated in deformed state to take into account possible bandgaps compressed into the same frequency range through higher modal branches.

As shown in Figs. 7.6b and 7.7, in the undeformed state the total frequency RBW of 1.06 is produced by four bandgaps in the range of $f_d = 0.16-0.56$ for extreme topology TQ1, and total frequency RBW of 0.66 is produced by four bandgaps in the range of $f_d = 0.19-0.53$ for extreme topology TQ10. Then in the deformed state according to Figs. 7.6c and 7.7, the total RBW of TA1 is dropped to 0.51 through two bandgaps in the range of $f_d = 0.24-0.51$ and bandgaps of TP10 are totally closed. Hence the later optimisation which is performed based on modified Tunability Objective-2, leads to an improved set of Pareto topologies compared to results of Tunability Objective-1. Wider initial bandgaps are evident over a slightly wider frequency range and extreme topology TB10 absolutely closes by prescribed deformation. Furthermore, the bandgap of deformed state over topologies TQ1 to TQ10 is mainly produced by a single gap which has considerably lower magnitude over the Pareto front (Fig. 7.6c) as compared to topologies

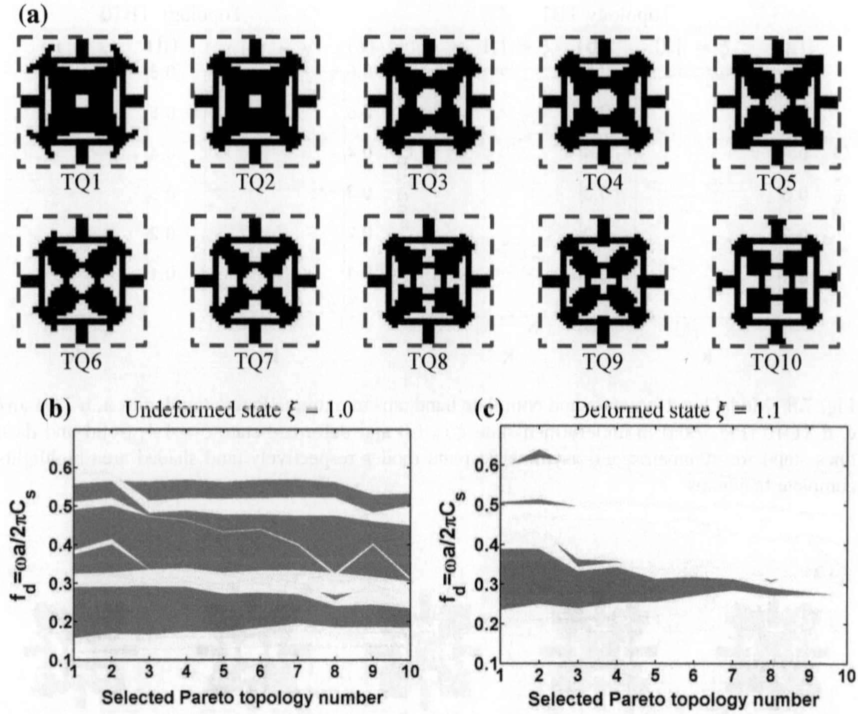

Fig. 7.6 **a** Selected Pareto topologies optimised for Tunability Objective-2; **b** bandgap gradient versus Pareto topology number in undeformed state $\xi = 1.0$ and **c** in deformed state $\xi = 1.1$

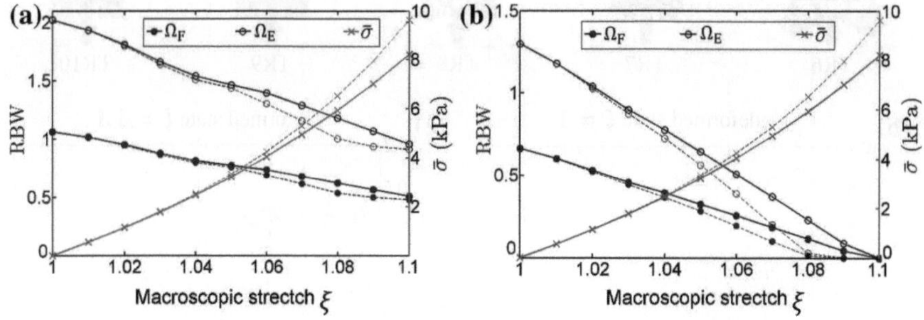

Fig. 7.7 Gradient of total RBW of first 12 plate modes and average macroscopic tuning stress $\bar{\sigma}$, versus macroscopic stretch of up to $\xi = 1.1$ for extreme Pareto topologies **a** TQ1 and **b** TQ10 (Fig. 7.6a); solid lines for assumed $J_m = 0.5$ and dash lines for additional $J_m = 0.1$

TP1 to TP10 (Fig. 7.3c). In fact, for improved Tunability Objective-2 the part of bandgap in the deformed state that exceeds the maximum bandgap frequency of the undeformed state, if any, is neglected. Therefore the optimisation is more focused on closing lower bandgaps.

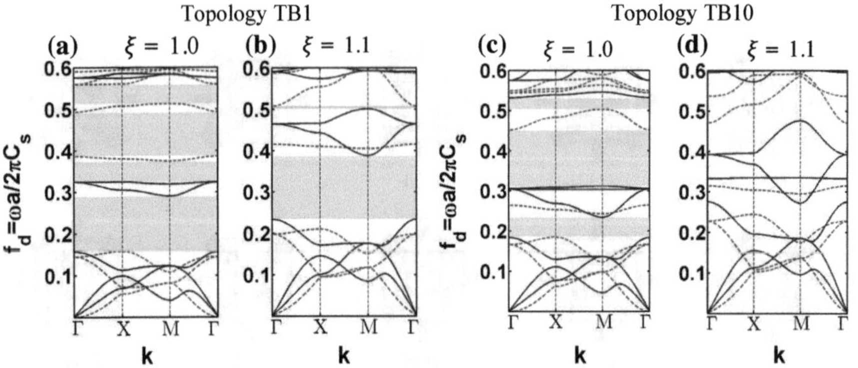

Fig. 7.8 Modal band structure and complete bandgaps of extreme Pareto topologies **a**, **b** TQ1 and **c**, **d** TQ10 (Fig. 7.6a) in undeformed state $\xi = 1.0$ and deformed state $\xi = 1.1$; Solid and dash lines stand for symmetric and asymmetric plate modes respectively, and shaded area highlights complete bandgaps

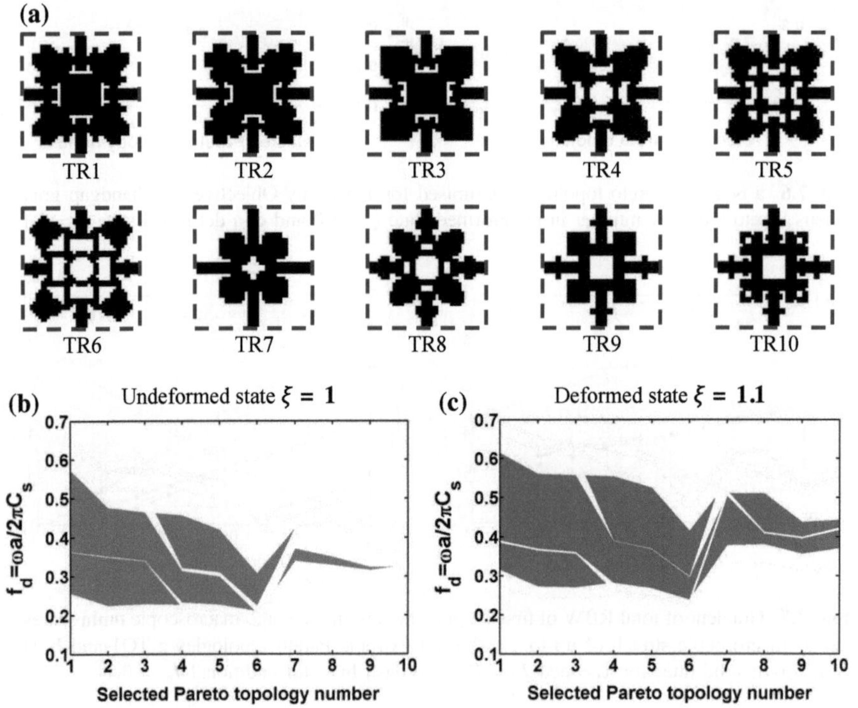

Fig. 7.9 **a** Selected Pareto topologies optimised for Tunability Objective-3; **b** bandgap gradient versus Pareto topology number in undeformed state $\xi = 1$ and **c** in deformed state $\xi = 1.1$

Ideally, the Pareto extreme for the first objective concerned with widest initial bandgap should be the same for both objectives. However, the differences between tunability objectives biasedly decline the Pareto population. Also, the optimisation has been performed twice for different objectives and may have been affected by random nature of GA algorithm to some extent. These deficiencies may be overcome by larger population size.

As shown in Fig. 7.8, similar to the results of Tunability Objective-1, modal branches are shifted up by applied stretch and modal bandgaps are predominantly closed by expanded symmetric plate modes mainly in second and third Brillouin zone borders XM and MΓ. Based on average tuning stresses given in Figs. 7.4 and 7.7, the required stress for deformation of extreme PhP unit-cells of both objectives is very low and almost equivalent at approximately $\bar{\sigma} = 8$ kPa. Hence the required load to stretch 1 cm^2 of lattice boundary area is just 0.8 N.

As mentioned earlier, in order to explore the effect of material nonlinearity on bandgap tunability efficiency, extreme topologies TQ1 and TQ0–10 are remodelled by additional elastomeric material of $J_m = 0.1$ and relevant results (dash lines) are included in Fig. 7.7. Accordingly, as expected, this material with higher stress stiffening nature leads to increased tuning stress of $\bar{\sigma} = 9.6$ kPa and improved bandgap tunability. Although the bandgaps of initial and deformed states are negligibly affected, higher bandgap degradation at lower levels of applied stretch is evident.

Pareto front solutions presented in this section introduce novel PhP unit-cells with degraded bandgap through deformation. Considering Prato topologies TQ1 to TQ10 (Fig. 7.6), if complete degradation of all bandgaps through prescribed deformation is desirable, then the extreme topology TQ10 would be the best choice. However, higher bandgaps of other topologies particularly TQ5 to TQ9 close through prescribed deformation. Moreover, the number and relative width of bandgaps vary from one topology to another. So, the supremacy of obtained Pareto topologies depends on intended application and desired operational frequency range.

7.4.2 Minimised RBW Gradient

The Tunability Objective-3 and its modified Tunability Objective-4 aim to maximise the initial bandgap efficiency and minimise the value of relative RBW gradient induced by prescribed stretch. The tunability objective function could be minimised by minimising the numerator (i.e. zero RBW gradient) on the positive side, and/or by minimising the denominator and negative numerator (i.e. minimise undeformed RBW and maximise deformed RBW) leading to a large negative value. Consequently, the obtained Pareto fronts are comprised of two distinct domains. One domain is converged to topologies with relatively wide bandgaps in the undeformed state having minimised sensitivity to the deformation. But the other domain contains topologies with negative deformation-induced bandgap gradient as they have small to negligible bandgaps in the undeformed state and widened or opened bandgaps in the deformed state. These novel topologies could be used in different applications: (i) widest bandgap with minimum sensitivity to external load, (ii) deformation-induced bandgaps which do not exist in initial state, and (iii) intermediate topologies with initial bandgap efficiency widened via external load.

The selected Pareto topologies of Tunability Objective-1, TR1 to TR10, and relevant complete bandgaps before and after deformation are shown in Fig. 7.9. Variation of RBW and average tuning stress versus stretch for extreme topologies are also presented in Fig. 7.10. The extreme Pareto topology TR1 has the widest initial frequency RBW of 0.79 over the frequency range of $f_d = 0.26-0.57$, which slightly reduces to 0.77 over $f_d = 0.32-0.62$. On the other side of Pareto front the extreme topology TR10 with no initial bandgap, shows frequency RBW of 0.17 by two bandgaps over frequency range of $f_d = 0.37-0.44$. As per modal band structures of these two extreme topologies presented in Fig. 7.11, all bandgaps are marginally confined by both symmetric and asymmetric plate modes at lower and upper sides and divided by an interrupting asymmetric plate mode.

The modified Tunability Objective-4, however leads to a different set of Pareto topologies TS1 to TS10 as presented and evaluated in Figs. 7.12, 7.13, and 7.14. The extreme topology TS1 has the highest undeformed frequency RBW of 0.97 over the range of $f_d = 0.18-0.50$ with minimal deformation-induced reduction to 0.84 over $f_d = 0.24-0.55$. Although the bandgap width is nearly maintained during deformation, its frequency range is shifted up leading to lower RBW. On the other side of Pareto front, the selected extreme topology TS9 has almost no initial bandgap (around 0.01) leading to a frequency RBW of 0.22 over $f_d = 0.43-0.53$. Topology TS9 has wider deformation-induced bandgap compared with topology TS10 but at higher frequency range.

According to the modal band structures shown in Fig. 7.14, the extreme topology TS1 in undeformed state includes two adjacent bandgaps interrupted by an asymmetric mode and confined by both symmetric and asymmetric mode types

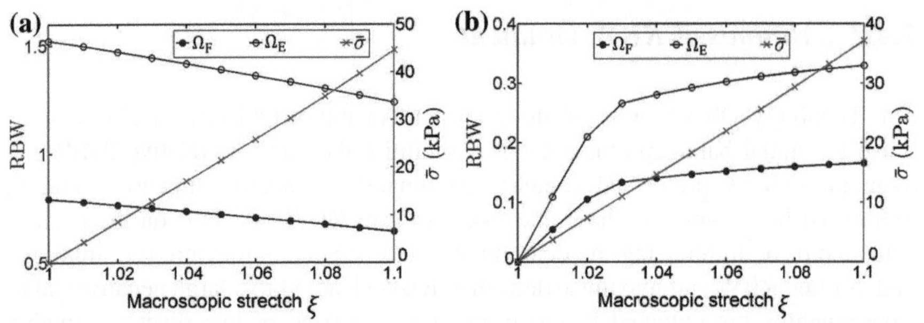

Fig. 7.10 Gradient of total RBW of first 12 plate modes and average macroscopic tuning stress $\bar{\sigma}$, versus macroscopic stretch of up to $\xi = 1.1$ for extreme Pareto topologies **a** TR1 and **b** TR10 (Fig. 7.9a)

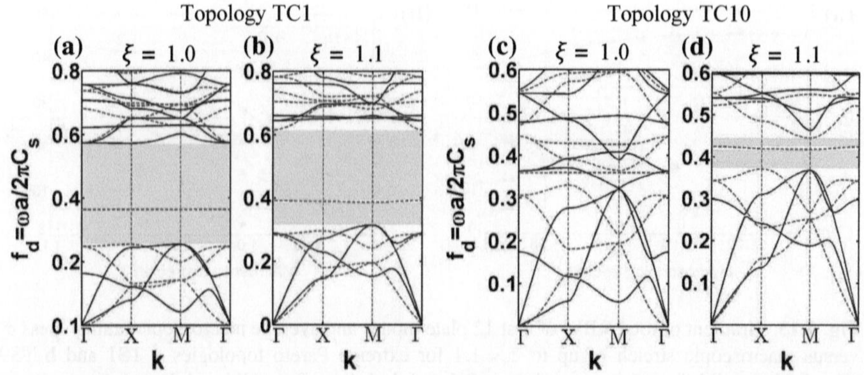

Fig. 7.11 Modal band structure and complete bandgaps of extreme Pareto topologies **a**, **b** TR1 and **c**, **d** TR10 (Fig. 7.9a) in undeformed state $\xi = 1$ and deformed state $\xi = 1.1$; Solid and dash lines stand for symmetric and asymmetric plate modes respectively, and shaded area highlights complete bandgaps

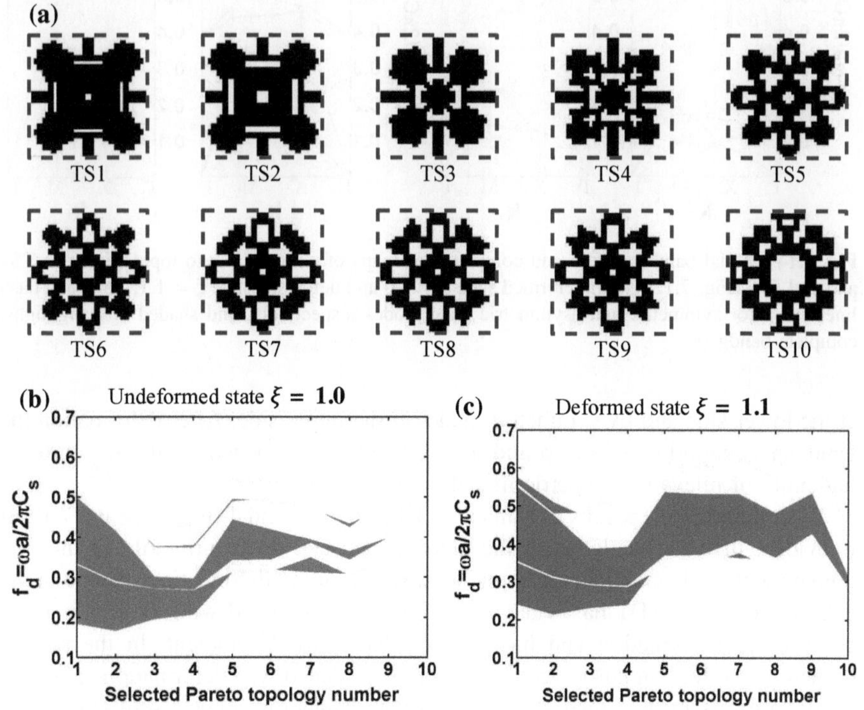

Fig. 7.12 **a** Selected Pareto topologies optimised for Tunability Objective-4; **b** bandgap gradient versus Pareto topology number in undeformed state $\xi = 1$ and **c** in deformed state $\xi = 1.1$

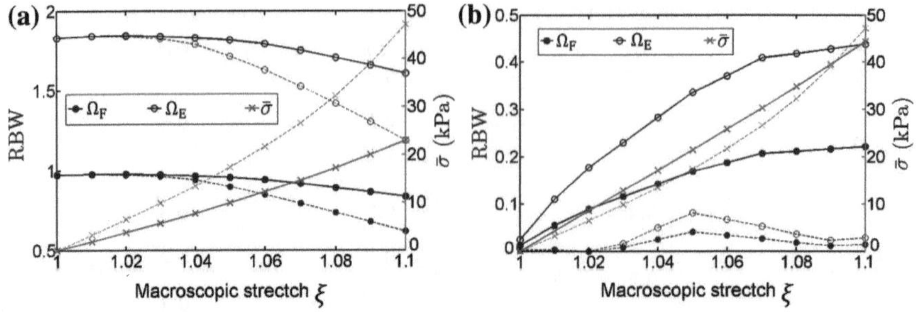

Fig. 7.13 Gradient of total RBW of first 12 plate modes and average macroscopic tuning stress $\bar{\sigma}$, versus macroscopic stretch of up to $\xi = 1.1$ for extreme Pareto topologies **a** TS1 and **b** TS9 (Fig. 7.12a); solid lines for assumed $J_m = 0.5$ and dash lines for additional $J_m = 0.1$

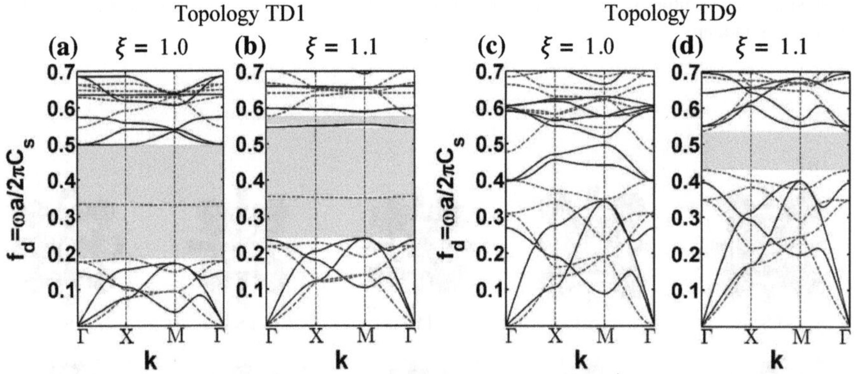

Fig. 7.14 Modal band structure and complete bandgaps of extreme Pareto topologies **a, b** TS1 and **c, d** TS9 (Fig. 7.12a) in undeformed state $\xi = 1.0$ and deformed state $\xi = 1.1$; Solid and dash lines stand for symmetric and asymmetric plate modes respectively, and shaded area highlights complete bandgaps

at the lower side and by symmetric modes at the upper side. After deformation the bandgap is slightly shifted up and a new bandgap is opened on the top through flattening of relevant symmetric modal branches.

By comparing the results of Tunability Objectives-3 and 4 (Figs. 7.9 and 7.12) it is evident that the bandgaps of the latter are pushed down primarily in the first domain with wide initial bandgap and least sensitivity to deformation. Although the extreme topology TD1 has wider initial bandgap compared with TC1, it is more sensitive to deformation and has higher relative RBW gradient. In the second domain regarding initially narrow bandgaps widened via deformation, a single bandgap exists in deformed state for modified Tunability Objective-4 (Fig. 7.9c) while that of Tunability Objective-3 is divided by an interrupting mode (Fig. 7.12c).

The average macroscopic tuning stresses $\bar{\sigma}$ given in Figs. 7.10 and 7.13 represent considerably higher stiffness topologies obtained for minimised tunability objectives as compared with the topologies corresponding to the tunability objective presented in Sect. 4.1. The extreme topology TC1 with maximum $\bar{\sigma} =$ 44.6 kPa is much stiffer than TD1 with maximum $\bar{\sigma} = 18.5$ kPa leading to lower initial bandgap as well as lower sensitivity to deformation for TC1.

On the other hand, the topology TS9 with $\bar{\sigma} = 44.3$ kPa is stiffer than TR10 with $\bar{\sigma} = 37.1$ kPa and provides higher deformation-induced RBW compared with TR10. So in order to open a frequency RBW of 0.22 through TS9 a tuning load of 4.43 N is required per 1 cm^2 of lattice boundary area, exceeding five times the tuning load already obtained for closing initial frequency RBW of 0.68 through TQ1. The results concerning additional elastomeric material with $J_m = 0.1$ included in Fig. 7.13 explain the tunability performance of extreme topologies TS1 and TS9 under higher stress stiffening condition. The compliant topology TS1 with $J_m = 0.1$ has much higher tuning load leading to considerably higher bandgap gradient compared with that of $J_m = 0.5$.

The stiff topology TS9 however shows slightly increased tuning stress but completely degraded bandgap tunability for $J_m = 0.1$.

So, after replacing constitutive material with a new one of higher stress stiffening properties, none of selected topologies performs as desired. One would of course have to optimise using these new material properties, potentially leading to new topologies that offer enhanced performance.

The same argument already stated about differences of Tunability Objectives-1 and 2 applies to Tunability Objectives-3 and 4. The deformation-induced bandgap of individual designs exceeding maximum bandgap frequency of initial state, if any, is disregarded in modified objective-4 which focuses the optimisation on bandgaps of lower bands. Ideally, the Pareto extreme for the first objective concerned with widest initial bandgap should be the same for all four objectives. However, the differences between tunability objectives biasedly decline the Pareto population. Also, the optimisation has been performed four times for different objectives and may have been affected by random nature of GA algorithm. These deficiencies may be overcome by larger population size.

As expected, the results for minimised RBW gradient introduced novel PhP unit-cells with different tunability characteristics. In the results concerning modified Tunability Objective-4, the first extreme topology TR1 has maximised bandgap efficiency in undeformed state and minimised sensitivity to deformation (e.g. caused by unwanted disturbances or operational loads). The other extreme topologies TR9 or TR10 with no bandgap in undeformed state, show opened bandgap through applied deformation. An intermediate topology like TR5 is a good choice with considerable initial bandgap whose bandgap efficiency is enhanced by applied external load.

7.5 Frequency Spectrum of Finite Tunable PhP Lattices

The bandgap efficiency and tunability performance of selected optimised Pareto topologies were demonstrated in the previous section. The analysis was performed on PhP unit-cells with Bloch-Floquet periodic boundary conditions and under the assumption of infinite periodicity. The aim of this section is to examine the frequency response of finite PhP lattices and present the performance of optimised topologies with finite periodicity. For this purpose, 7 × 8 rectangular lattice arrays of Pareto topologies TQ10 (Fig. 7.15d) and TS9 (Fig. 7.16d) resulting from

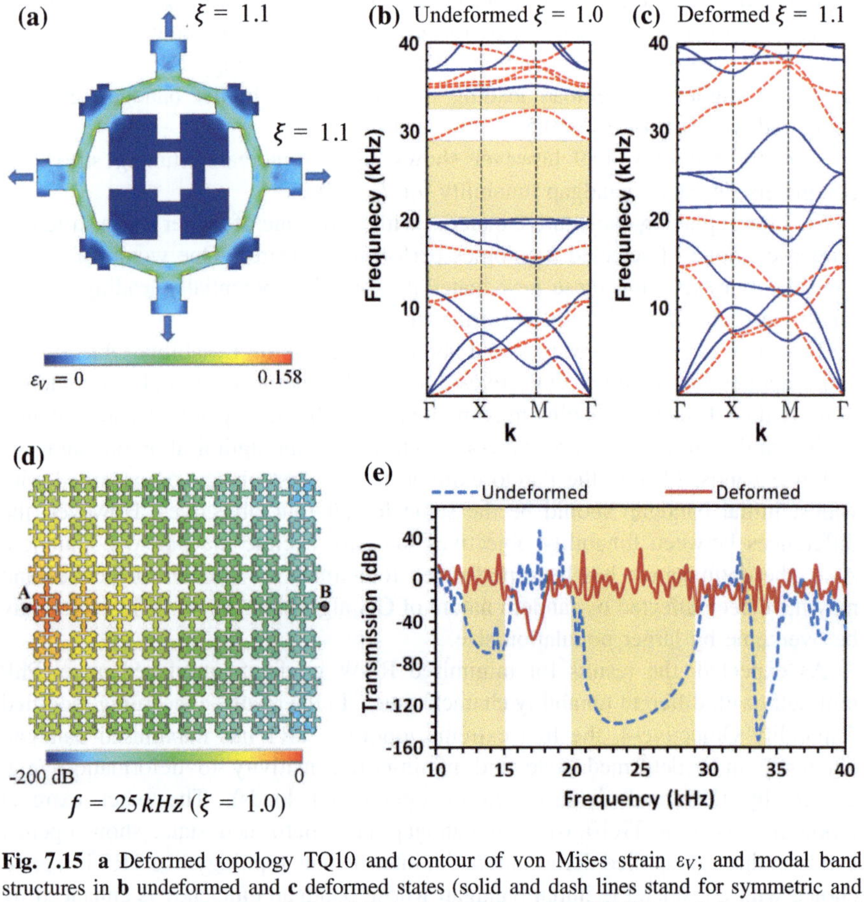

Fig. 7.15 **a** Deformed topology TQ10 and contour of von Mises strain ε_V; and modal band structures in **b** undeformed and **c** deformed states (solid and dash lines stand for symmetric and asymmetric plate modes respectively); **d** Finite lattice of topology TQ10 in undeformed state and relevant steady state transmission contour at bandgap frequency 25 kHz; **e** steady state transmission spectrum of guided waves to point B in undeformed and deformed states stimulated at point A, (Unit-cell width $a = 0.5$ mm and height $h = 0.25$ mm)

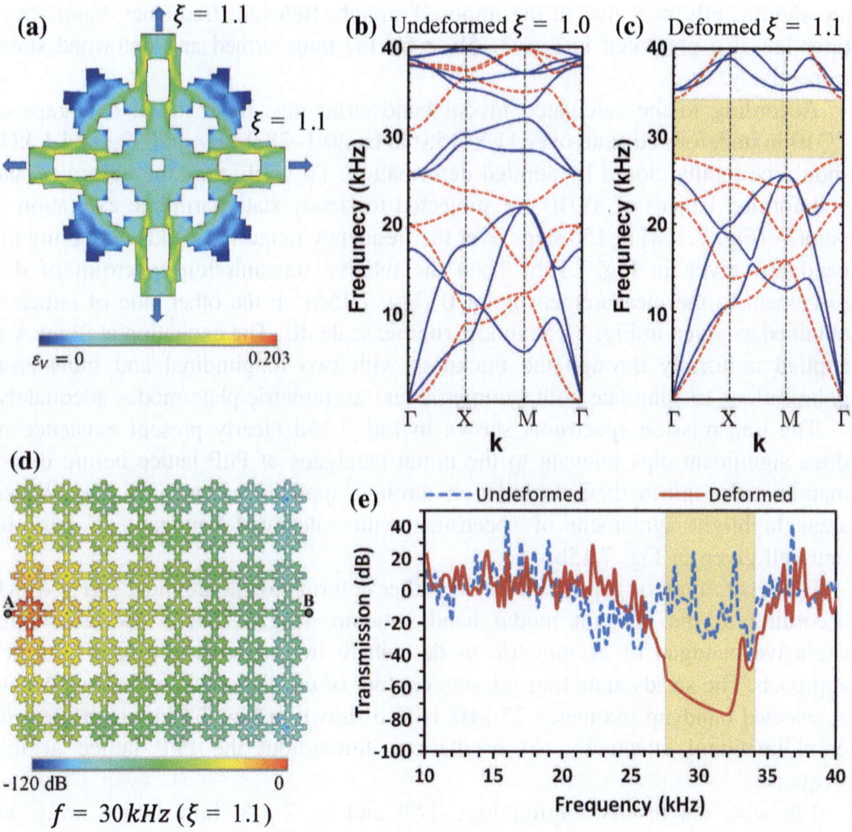

Fig. 7.16 **a** Deformed topology TS9 and contour of von Mises strain ε_V; and modal band structures in **b** undeformed and **c** deformed states (solid and dash lines stand for symmetric and asymmetric plate modes respectively); **d** Finite lattice of topology TS9 in deformed state and relevant steady state transmission contour at bandgap frequency 30 kHz; **e** steady state transmission spectrum of guided waves to point B in undeformed and deformed states stimulated at point A, (Unit-cell width $a = 0.5$ mm and height $h = 0.25$ mm)

Tunability Objectives-2 and 4 are modelled by a fine mesh of ANSYS quadratic elements Solid186. The frequency response under steady state excitation is calculated and the transmission spectrum of excited wave into a selected point of structure is defined in logarithmic scale dB, in both undeformed and deformed states. Unit-cells of width $a = 0.5$ mm and height $h = 0.25$ mm are modelled producing phononic controllability over 10–35 kHz.

The deformed geometry and contour of von Mises strain ε_V of topology TQ10 under periodic boundary conditions prescribing macromechanical stretch of $\lambda = 1.1$ are shown in Fig. 7.15a. Accordingly the central segment of topology with diagonal connections to initially square frame is almost strain free and all edges of this frame are significantly bent due to the imposed stretch. Relevant frequency band structures are also presented in Fig. 7.15b, c for the undeformed and deformed states respectively.

According to the calculated modal band structures, three initial bandgaps of TQ10 in undeformed state over 11.8–15.0 kHz, 20.1–28.9 kHz and 32.1–34.1 kHz should be totally closed by applied deformation. To verify this, the deformed and undeformed lattices of TQ10 are subjected to steady state harmonic excitation at point A (Fig. 7.15d) in 150 steps over the frequency range 10–40 kHz covering the bandgaps given in Fig. 7.15b. Then the relative transmission spectrum of displacement to the measurement point B (Fig. 7.15d) on the other side of lattice is obtained as given in Fig. 7.15e in logarithmic scale dB. The excitation at point A is applied uniformly through the thickness with two longitudinal and transversal polarisations to stimulate both symmetric and asymmetric plate modes adequately.

The transmission spectrums shown in Fig. 7.15d clearly present existence of three significant dips relevant to the initial bandgaps of PhP lattice before deformation and confirm their degradation through applied deformation. The shaded areas highlight agreement of spectrums with calculated bandgaps of periodic unit-cell given in Fig. 7.15b.

However, a minor gap is still evident after deformation at around 17 kHz which according to the relevant modal band structure (Fig. 7.15c) is located within exclusive bandgap of asymmetric modes where two symmetric modal branches approach. The steady state transmission contour of oscillations in undeformed state at selected bandgap frequency 25 kHz is also shown in Fig. 7.15d which presents omnidirectional attenuation of oscillations throughout the PhP lattice at this frequency.

Likewise the results for topology TS9 and its 7×8 finite lattice array are compared in Fig. 7.16. Deformed unit-cell under periodic boundary condition and relevant von Mises strain contour as shown in Fig. 7.16a reveal how the central inclusion with orthogonal connections to its surrounding frame resists to deformation leading to higher stiffness compared with TQ10. For the same macromechanical stretch of $\lambda = 1.1$ the maximum micromechanical strain in the unit-cell of TS9 (Fig. 7.16a) $\varepsilon_v = 0.203$ is higher than that of TQ10 (Fig. 7.15a) $\varepsilon_v = 0.158$.

As predicted by modal band structures of undeformed and deformed states concerning topology TR9 shown in Fig. 7.16c and d respectively, a complete bandgap is induced by deformation over frequency range of 27.6–34.0 kHz. This is clearly observable in Fig. 7.16e by the relevant dip in transmission spectrum of oscillations in the finite lattice from the excitation point A to the measurement point B. The modal band structure of undeformed state in Fig. 7.16b shows a tiny gap at around 22 kHz between approaching modal branches which has caused a minor attenuation in spectrum of undeformed state around this frequency (Fig. 7.16e). The transmission contour of oscillations throughout the deformed lattice of TR9 as shown in Fig. 7.16d confirms the omnidirectional decay of wave at selected bandgap frequency 30 kHz.

7.6 Concluding Remarks

In this research, novel porous phononic plate (PhP) topologies with optimised bandgap tunability performance were achieved through a multi-objective optimisation strategy. The deformation-induced tunability of complete bandgaps of guided waves within the first 12 modal branches was investigated. Macromechanical deformation of phononic lattice under prescribed equibiaxial stretch of 10% was assumed for this purpose. Maximum relative bandgap width (RBW) of guided waves in the undeformed state and maximum deformation-induced bandgap gradient were the two objectives of optimisation. NSGA-II was adopted as a genetic algorithm based optimisation strategy for this two objective, multimodal-multistate optimisation problem.

Different tunability targets were introduced to enhance or degrade bandgap efficiency via deformation, and modified tunability objectives were examined by considering the initial bandgap frequency range (in the undeformed state) as a reference for calculation of bandgap gradient. The resultant Pareto fronts provide broad distribution of PhP topologies with varying bandgap efficiency and bandgap tunability characteristics. Originally wide bandgap topologies with minimum sensitivity or maximum sensitivity (in either ascending or descending fashion) to the prescribed tuning load were defined. Following conclusions can be made from the results of this part of study:

- The topologies obtained in this study, demonstrate new PhP designs with promising application in active control (i.e. filtration, resonation and steering) of particular guided wave modes, and maximised/minimised sensitivity of bandgap efficiency to external excitations/disturbances.
- Pareto front solutions obtained for maximised RBW gradient introduce novel PhP unit-cells with degraded bandgap through deformation. All optimised topologies have maximised sensitivity to the prescribed deformation. An extreme topology has highest bandgap efficiency but not totally degraded through deformation. Another extreme topology has lower bandgap efficiency which completely closes through deformation.
- The results for minimised RBW gradient also introduce novel PhP unit-cells with other promising tunability characteristics: (i) widest bandgaps with minimum sensitivity to external load, (ii) deformation-induced bandgaps which do not exist in initial state, and (iii) intermediate topologies with initial bandgap efficiency widened via external load
- The modified tunability objectives were shown to provide totally different and generally better Pareto solutions. For improved tunability objective, the part of bandgap in the deformed state that exceeds the maximum bandgap frequency of the undeformed state, if any, is neglected. Therefore the optimisation is more focused on closing lower bandgaps.
- The topologies are optimised based on a relatively low stress stiffening with $J_m = 0.5$. So a few optimised topologies were reanalysed for higher stress stiffening with $J_m = 0.1$. This higher material nonlinearity enhanced the

deformation-induced tunability of topologies with maximised RBW gradient. In contrary, the tunability of topologies with minimised RBW gradient was degraded by this higher material nonlinearity. Hence, in the latter case new optimisation is essential for other values of J_m.

- The bandgap and tunability performance of finite lattices made by selected topologies with deformation-induced closed and opened bandgaps were in good agreement with predicted modal band structures of relevant unit-cells under perfect periodicity.

References

Bayat, A., & Gordaninejad, F. (2015a). Switching band-gaps of a phononic crystal slab by surface instability. *Smart Materials and Structures, 24*(7), 075009.

Bayat, A., & Gordaninejad, F. (2015b). Dynamic response of a tunable phononic crystal under applied mechanical and magnetic loadings. *Smart Materials and Structures, 24*(6), 065027.

Bergamini, A., Delpero, T., Simoni, L. D., Lillo, L. D., Ruzzene, M., & Ermanni, P. (2014). Phononic crystal with adaptive connectivity. *Advanced Materials, 26*(9), 1343–1347.

Bertoldi, K., & Boyce, M. C. (2008). Wave propagation and instabilities in monolithic and periodically structured elastomeric materials undergoing large deformations. *Physical Review B, 78*(18), 184107.

Bertoldi, K., Boyce, M. C., Deschanel, S., Prange, S. M., & Mullin, T. (2008). Mechanics of deformation-triggered pattern transformations and superelastic behavior in periodic elastomeric structures. *Journal of the Mechanics and Physics of Solids, 56*(8), 2642–2668.

Bilal, O. R., & Hussein, M. I. (2012). Topologically evolved phononic material: Breaking the world record in band gap size. In *Photonic and phononic properties of engineered nanostructures* (pp. 826911–826917). International Society for Optics and Photonics.

Chapman, C. D., Saitou, K., & Jakiela, M. J. (1994). Genetic algorithms as an approach to configuration and topology design. *Journal of Mechanical Design, 116*(4), 1005–1012.

de Borst, R., Crisfield, M. A., Remmers, J. J. C., & Verhoosel, C. V. (2012). *Non-linear finite element analysis of solids and structures* (2nd ed.). New York: Wiley.

Dong, H.-W., Su, X.-X., & Wang, Y.-S. (2014a). Multi-objective optimization of two-dimensional porous phononic crystals. *Journal of Physics. D. Applied Physics, 47*(15), 155301.

Dong, H.-W., Su, X.-X., Wang, Y.-S., & Zhang, C. (2014b). Topological optimization of two-dimensional phononic crystals based on the finite element method and genetic algorithm. *Structural and Multidisciplinary Optimization, 50*(4), 593–604.

Dong, H.-W., Su, X.-X., Wang, Y.-S., & Zhang, C. (2014c). Topology optimization of two-dimensional asymmetrical phononic crystals. *Physics Letters A, 378*(4), 434–441.

Evgrafov, A., Rupp, C. J., Dunn, M. L., & Maute, K. (2008). Optimal synthesis of tunable elastic wave-guides. *Computer Methods in Applied Mechanics and Engineering, 198*(2), 292–301.

Filipovic, N., Stojanovic, B., Kojic, N., & Kojic, M. (2008). *Computer modeling in bioengineering-theoretical background. Examples and Software*. New York: Wiley.

Gazonas, G. A., Weile, D. S., Wildman, R., & Mohan, A. (2006). Genetic algorithm optimization of phononic bandgap structures. *International Journal of Solids and Structures, 43*(18–19), 5851–5866.

Gei, M., Bigoni, D., Movchan, A., & Bacca, M. (2013). *'Band-gap properties of prestressed structures', Acoustic Metamaterials* (pp. 61–82). Berlin: Springer.

Goffaux, C., & Vigneron, J. P. (2001, July 31). Theoretical study of a tunable phononic band gap system. *Physical Review B, 64*(7), 075118.

Guest, J. K., Prévost, J., & Belytschko, T. (2004). Achieving minimum length scale in topology optimization using nodal design variables and projection functions. *International Journal for Numerical Methods in Engineering, 61*(2), 238–254.

Guest, J. K., & Smith Genut, L. C. (2010). Reducing dimensionality in topology optimization using adaptive design variable fields. *International Journal for Numerical Methods in Engineering, 81*(8), 1019–1045.

Hajela, P., Lee, E., & Lin, C. Y. (1993). Genetic algorithms in structural topology optimization. In M. Bendsøe & C. M. Soares (Eds.), *Topology design of structures* (Vol. 227, pp. 117–133). Dordrecht: Springer Netherlands.

Hussein, M. I., Hamza, K., Hulbert, G. M., & Saitou, K. (2007, November). Optimal synthesis of 2D phononic crystals for broadband frequency isolation. *Waves in Random and Complex Media, 17*(4), 491–510.

Hussein, M. I., Hamza, K., Hulbert, G. M., Scott, R. A., & Saitou, K. (2006, January 1). Multiobjective evolutionary optimization of periodic layered materials for desired wave dispersion characteristics. *Structural and Multidisciplinary Optimization, 31*(1), pp. 60–75.

Lin, S. C. S., & Huang, T. J. (2011, May 18). Tunable phononic crystals with anisotropic inclusions. *Physical Review B, 83*(17), 174303.

Liu, Z., Wu, B., & He, C. (2014). Band-gap optimization of two-dimensional phononic crystals based on genetic algorithm and FPWE. In *Waves in Random and Complex Media*, no. ahead-of-print (pp. 1–20).

Manktelow, K. L., Leamy, M. J., & Ruzzene, M. (2013). Topology Design and Optimization of Nonlinear Periodic Materials. *Journal of the Mechanics and Physics of Solids, 61*(12), 2433–2453.

Matar, O. B., Vasseur, J., & Deymier, P. A. (2013). *'Tunable phononic crystals and metamaterials', acoustic metamaterials and phononic crystals* (pp. 253–280). Berlin: Springer.

Olsson Iii, R. H., & El-Kady, I. F. (2009). Microfabricated phononic crystal devices and applications. *Measurement Science & Technology, 20*(1), 012002.

Olsson Iii, R. H., El-Kady, I. F., Su, M. F., Tuck, M. R., & Fleming, J. G. (2008). Microfabricated VHF acoustic crystals and waveguides. *Sensors and Actuators A: Physical, 145–146*(0, 7), 87–93.

Pratap, A., Agarwal, S., & Meyarivan, T. (2002). A fast and elitist multiobjective genetic algorithm: NSGA-II. *IEEE Transactions on Evolutionary Computation, 6*(2), 182–197.

Rudykh, S., & Boyce, M. C. (2014). Transforming wave propagation in layered media via instability-induced interfacial wrinkling. *Physical Review Letters, 112*(3), 034301.

Rupp, C. J., Dunn, M. L., & Maute, K. (2010). Switchable phononic wave filtering, guiding, harvesting, and actuating in polarization-patterned piezoelectric solids. *Applied Physics Letters, 96*(11), 111902.

Sigmund, O., & Jensen, J. S. (2003). Systematic design of phononic band-gap materials and structures by topology optimization. *Philosophical Transactions of the Royal Society, 361* (2003), 1001–1019.

Wang, P., Casadei, F., Shan, S., Weaver, J. C., & Bertoldi, K. (2014). Harnessing buckling to design tunable locally resonant acoustic metamaterials. *Physical Review Letters, 113*(1), 014301.

Wang, P., Shim, J., & Bertoldi, K. (2013). Effects of geometric and material nonlinearities on tunable band gaps and low-frequency directionality of phononic crystals. *Physical Review B, 88*(1), 014304.

Weise, T. (2006). *Global optimization algorithms—Theory and application*, http://www.it-weise.de/.

Yao, Y., Wu, F., Zhang, X., & Hou, Z. (2011). Thermal tuning of Lamb wave band structure in a two-dimensional phononic crystal plate. *Journal of Applied Physics, 110*(12), 123503.

Chapter 8
Experimental Validation of Optimised Porous 2D PhPs

8.1 Introduction

In the preceding chapters topology optimisation study was carried out, and optimised unit-cells and their representative finite plates were computationally analysed and validated. However, it was desirable to manufacture selected optimised topologies and experimentally observe the agreement between their measured and calculated bandgap and stiffness properties.

For this purpose, critical stiff and compliant topologies of aspect ratio 10 from the results of Chap. 6 were manufactured by water-jetting of aluminium plates and laser cutting of Plexiglas (PMMA) plates. As the topology refinement significantly enhanced the supremacy of optimised topologies for bandgaps of asymmetric wave modes, both coarse and refined topologies were produced and their bandgap efficiencies were compared. For complete bandgaps of mixed wave modes selected coarse topologies are evaluated. The elastic modulus E_s, Poisson's ratio ν_s and density ρ_s of aluminium and PMMA samples are given in Table 8.1. Due to viscoelasticity of PMMA, its dynamic elastic modulus is measured and used for calculation of modal band structure.

8.2 Optimised PhPs for Exclusive Bandgap of Asymmetric Wave Modes

In this section the results concerning experimental evaluation of produced PhP topologies for bandgaps of asymmetric wave modes are presented and discussed. Aluminium PhPs of a few selected topologies (of both coarse and fine resolutions) are water-jetted and experimentally validated. For each topology resolution 3 topologies are chosen for production (Fig. 6.2): compliant topologies (TI4, TJ3)

S. Hedayatrasa, *Design Optimisation and Validation of Phononic Crystal Plates for Manipulation of Elastodynamic Guided Waves*, Springer Theses,
https://doi.org/10.1007/978-3-319-72959-6_8

Table 8.1 Mechanical properties of constitutive materials used for production of PhPs

Constitutive material	E_s (GPa)	v_s	ρ_s (kg/m^3)
Aluminium	70	0.34	2700
Plexiglas (PMMA)	6.5 (Dynamic)	0.34	1180

and stiff topologies (TI5, TJ6) adjacent to the extremes of discontinuous domain of Pareto fronts, and the stiffest topology (TI8, TJ8) taken from the stiff regime.

Although the refined design domain introduces superior topologies compared with the coarse design domain, it requires double accuracy (resolution) for manufacturing. Moreover, the refined topologies include narrower void features which require lower perforation beam diameter of the waterjet. To handle this, the coarse topologies and refined topologies are manufactured in an aluminium plate with thickness of 2.5 and 5 mm, respectively, resulting in unit-cell dimensions of 25 and 50 mm. The features, effective elastic properties and bandgap frequencies of topologies selected for production are summarised in Table 8.2.

Table 8.2 The calculated effective elastic properties and bandgap frequencies of coarse and refined Pareto topologies selected for production (Fig. 6.2) to be compared with experiment results and earlier works

Description	Compliant topology		Stiff topology		High stiffness topology		Halkjær et al. (2006)	Bilal and Hussein (2012)
Resolution	32 × 32	64 × 64	32 × 32	64 × 64	32 × 32	64 × 64	–	64 × 64
Topology	TC4	TR3	TC5	TR6	TC8	TR8	–	–
a/h (mm/mm)	25/2.5	50/5	25/2.5	50/5	25/2.5	50/5	11	11
$E_r \times 10^3$	0.543	0.634	2.419	18.088	23.017	26.054	–	–
$G_r \times 10^3$	0.675	0.662	0.991	6.228	20.314	29.306	–	–
v_e	0.9220	0.9198	0.6341	0.2654	0.5252	0.5897	–	–
v_f (%)	22.65	23.34	28.90	36.43	37.50	43.55	–	–
$f_{d,\min}$	0.0178	0.0167	0.0212	0.0232	0.0300	0.0285	–	–
$f_{d,\max}$	0.0273	0.0281	0.0311	0.0369	0.0358	0.0366	–	–
Δf_d	0.0095	0.0114	0.0099	0.0137	0.0058	0.0081	–	–
$\bar{f}_d$	0.0226	0.0224	0.0262	0.0300	0.0329	0.0326	–	–
$\Delta f_d^2/\bar{f}_d^2$	0.8068	0.9560	0.7310	0.8668	0.3499	0.4901	–	–
$\Delta f_d/\bar{f}_d$	0.4213	0.5089	0.3786	0.4559	0.1763	0.2488	–	–
$f_{\min}$ (kHz)	3.63	1.71	4.31	2.36	6.10	2.90	–	–
$f_{\max}$ (kHz)	5.55	2.86	6.33	3.76	7.29	3.73	–	–
$\Delta f_d^2/\bar{f}_d^{2}$[a]	0.7449	0.8998	0.6790	0.8300	0.3195	0.4682	–	0.5639
$\Delta f_d/\bar{f}_d$[a]	0.3863	0.4753	0.3499	0.4346	0.1608	0.2374	0.16	–

[a]Calculated based on aspect ratio $a/h = 11$ and Polycarbonate solid background to compare with that of (Halkjær et al. 2006) and (Bilal and Hussein 2012)

To compare with experimental results, these topologies are analysed by a very fine FEM model using linear solid elements with in-plane mesh resolution of 256 × 256 and 20 element layers through thickness. This higher resolution reduced both calculated modal frequencies and homogenised elastic properties by less than 4% compared with resolution of 64 × 64 with 8 layers. The dimensionless frequencies of upper and lower limits of bandgap $f_{d,\min}$ and $f_{d,\max}$, respectively, bandgap width Δf_d, midgap frequency $\overline{f_d}$ and filling fraction v_f are given. The actual upper bandgap frequency $f_{\max}$ and lower bandgap frequency $f_{\min}$ of produced aluminum PhPs are also included.

In order to compare the bandgap efficiency of selected topologies with results of other researchers the RBW is calculated based on both normalised formulations $\Delta f_d^2/\overline{f_d^2}$ and $\Delta f_d/\overline{f_d}$. The topologies are then reanalysed by the assumption of aspect ratio $a/h = 11$ and Polycarbonate solid background (Bilal and Hussein 2012; Halkjær et al. 2006), and the relevant normalised bandgaps are included in Table 8.2 (marked by [a]). It is obvious that the increased aspect ratio and new material properties slightly reduce the RBW. For the post-processed optimised topology of Halkjær et al. (2006) RBW of $\Delta f_d/\overline{f_d} = 0.16$, and for the optimised topology of Bilal and Hussein (2012) RBW of $\Delta f_d^2/\overline{f_d^2} = 0.5639$ was reported, confirming the robustness of present work in obtaining a wide range of topologies with considerably higher bandgap efficiency.

8.2.1 Validation of Bandgap of Selected Coarse Topologies

For production of coarse topologies, rectangular aluminium plate of size 500 × 400 mm and thickness 2.5 mm was water-jetted to make a central lattice array of 10 × 8 PhP unit-cells with lattice periodicity of 25 mm as schematically shown in Fig. 8.1a for topology TI4.

In order to evaluate the bandgap efficiency of perforated section, the wave transmission from excitation points E_1 or E_2 on the uniform boundary to the measurement point R_2 is determined. The transmission of wave through uniform plate and its resonation by PhP section is also measured at points R_1 and R_3 for excitations at E_1 and E_2, respectively. Slots are cut on the border of designed PhP plate to isolate excitation segments and preferably prevent the excitation to reach the whole uniform border.

The steady state transmission contour of harmonic excitation at E_1 at bandgap frequency 5.5 kHz for PhP of topology TI5 is presented in Fig. 8.1b in logarithmic scale dB. The exponential wave attenuation throughout the PhP lattice, wave resonation at excitation segment and also the function of slots in preventing transmission of wave to the whole border is well illustrated in this contour.

Water-jetting equipment is shown in Fig. 8.2a, b, and close view of water-jetted topologies TI4, TI5 and TI8 are shown in Fig. 8.2c–e respectively. The water-jetted PhP specimen of topology TI8 is also shown in Fig. 8.2f. For excitation and

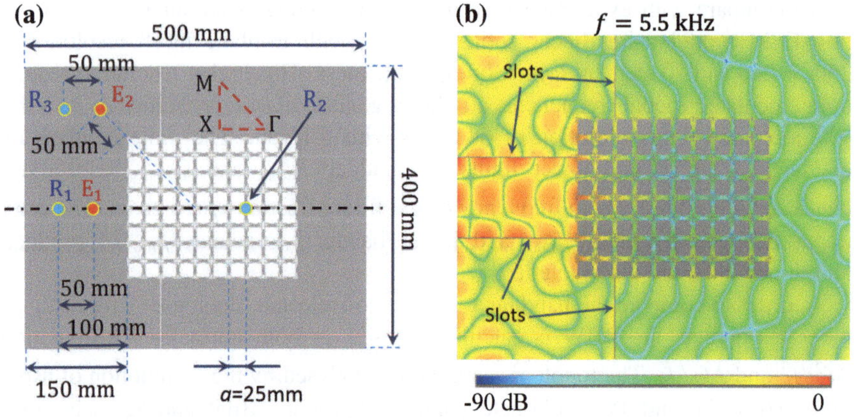

Fig. 8.1 **a** PhP plate design (shown for topology TI5) and relative location of excitation points E_1, E_2 and E_3 and measurement points R_1, R_2 and R_3, and representation of irreducible Brillouin zone ΓXM; **b** steady state displacement contour at bandgap frequency $f = 5.5$ kHz of topology TI4 applied on excitation point E_1

measurement of the flexural waves, circular PZTs are bonded to the top surface of plates with Phenyl Salicylate (Salol). PZTs with diameter of 20 mm are used for excitation and PZTs with diameter of 10 mm are used for measuring transmitted signals (Fig. 8.2f). Bigger exciting PZTs are preferably used to increase the stimulated wave amplitude. The four corners of the rectangular PhP specimens are suspended on small pieces of foam to prevent coupling of wave to the test table.

It is known that the oscillations induced by such circular PZT on one side of the plate dominantly excites the asymmetric wave modes and does not excite symmetric modes to a measurable level within low frequency range of designed bandgaps (Giurgiutiu 2003). The measured signals are recorded with a high resolution 24-bit acquisition card (NI9234) at a sampling frequency of 50 kHz for adequate evaluation of attenuated wave amplitude.

The frequency spectra are then calculated using fast Fourier transform (FFT) and are averaged over 20 measurements in order to improve the signal-to-noise ratio. All spectra are normalised to the highest transmission amplitude measured at point R_2.

A logarithmic frequency sweep excitation voltage V is first applied with amplitude of $V_0 = 100$ V, starting from $f_0 = 100$ Hz to $f_1 = 30$ kHz over sweep time of $t_1 = 1$ s:

$$V = V_0 \cos(2\pi f_0 \beta^t t) \tag{8.1}$$

where the coefficient $\beta = (f_1/f_0)^{1/t_1}$. This logarithmic sweep enables adequate excitation of low frequencies compared with a linear sweep. For topology TI5 points E_1 and E_2 are separately excited and relevant spectrums are shown in

Fig. 8.3a–c and d–f, respectively, up to 15 kHz. The transmission spectrum to the measurement points is shown on the left side. The relative transmission spectrum is also shown on the right hand side which is actually the transmission spectrum of R_2 in PhP section normalised by relevant measured transmission spectrum of uniform section (R_1 for excitation on E_1, or R_3 for excitation on E_2). In fact, the excited spectrum is amplified by resonation of uniform border and wave reflections from PhP section within bandgap frequency. Hence, the calculated relative transmission gives a better presentation of attenuation level of any frequency with respect to its relevant excitation level.

The calculated modal band structure is also included in the centre for verification in which the symmetric and asymmetric modal branches are shown by solid and dash lines, respectively. The dot lines highlight the frequency levels corresponding to complete bandgap and also partial bandgap along Brillouin zone border ΓX (Fig. 8.3b, e). The spectra shown in Fig. 8.3 evidently confirm the attenuation of transmission up to −50 dB within calculated bandgap frequency range of topology TI5. Of course, the transmission starts to drop from lower frequency level around 2 kHz at which a partial bandgap opens along Brillouin zone border ΓX. This is more clearly observed in the relative transmission from E_1 (Fig. 8.3c) which is dominantly orthogonal and along ΓX. But in the relative transmission from E_2 (Fig. 8.3f) the transmission is less affected by this partial bandgap.

The transmission spectrums of frequency sweep excitation on E_1 for other coarse topologies TI4 and TI8 is also shown in Fig. 8.4 in the same manner. The compliant topology TI4 has wider complete bandgap attenuating the transmission up to −60 dB (Fig. 8.4c), and in contrary the stiff topology TI8 has narrower complete bandgap with less attenuation efficiency up to below −40 dB (Fig. 8.4f). In both cases the transmissions start to drop from lower frequency level corresponding to opening partial bandgap along ΓX.

Although the complete bandgap of TI8 is narrower than TI4, its partial bandgap widths along ΓX and ΓM are comparable with TI4, of course at higher frequency range. The transmission spectrum of R_1 (or R_3) in Figs. 8.3a, d and 8.4a, d confirms the resonation of spectrum within bandgap frequency range due to reflections from PhP section. Comparison of the modal band structures of coarse topologies show how the modal branches are shifted up by increasing stiffness.

A random Gaussian white noise is further applied to the excitation point E_1 of topologies TI4 and TI5 for simultaneous broadband excitation of frequencies. Then the time-frequency transmission spectrums are calculated through Gabor wavelet transform as shown in Fig. 8.5a, b and c, d for TI4 and TI5, respectively. Spectra are normalised to the highest transmission amplitude measured at point R_2 of relevant topology. The wavelet spectra clearly illustrate the steady state filtration of transmission to point R_2 within calculated bandgap frequencies and in contrast the resonated transmission to point R_1. It is also shown that the compliant topology mode TI4 has slightly higher attenuation level within bandgap (Fig. 8.5b) compared

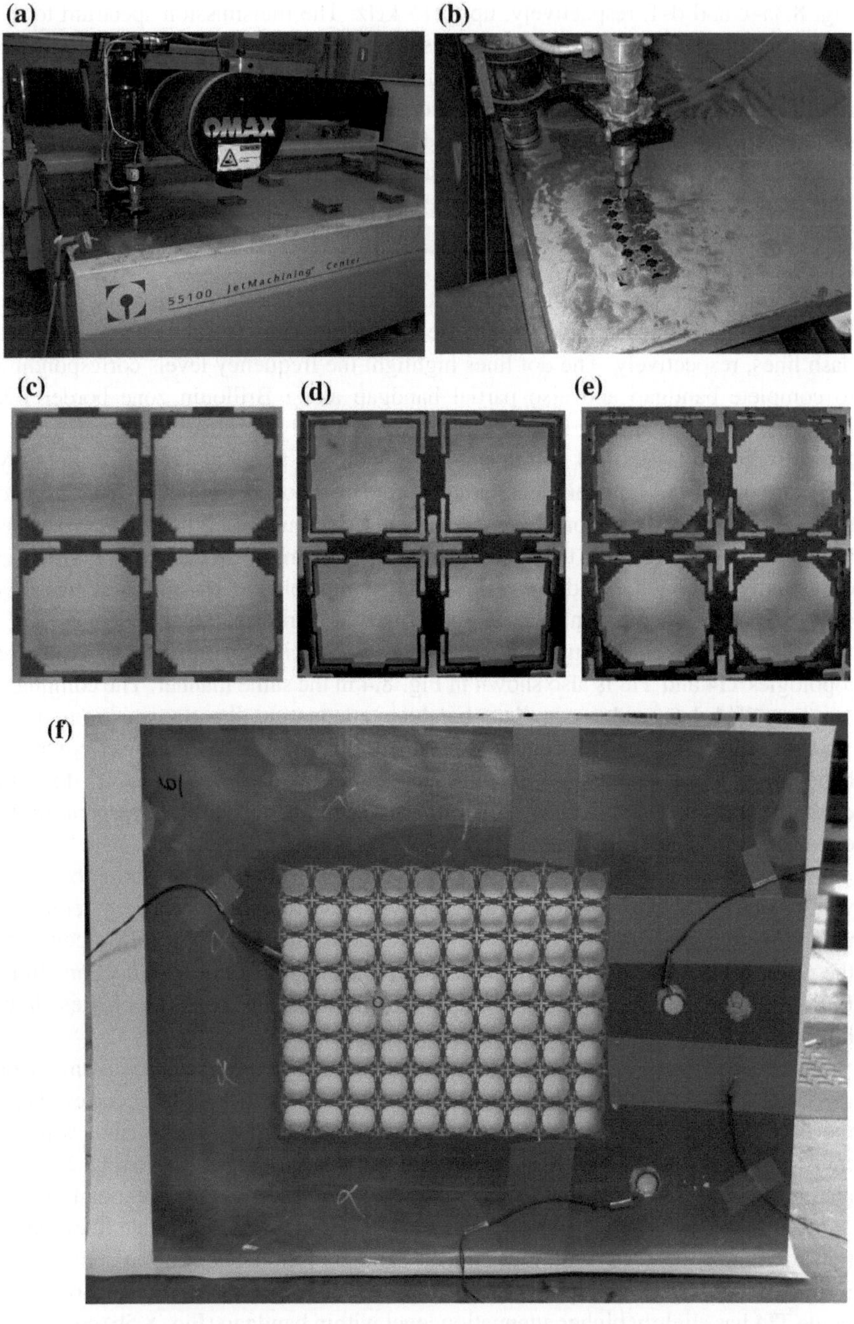
(a)
(b)
OMAX
55100 JetMachining® Center
(c)
(d)
(e)
(f)

◂**Fig. 8.2** **a**, **b** Water-jetting equipment, and water-jetted topology **c** TI4, **d** TI5 and **e** TI8 (plate thickness $h = 2.5$ mm and unit-cell width $a = 25$ mm), and **f** water-jetted PhP specimen of topology TI8 with attached PZTs (slots are covered by sticky tape to protect the cut specimen while carrying), (MMS Group, Ghent University)

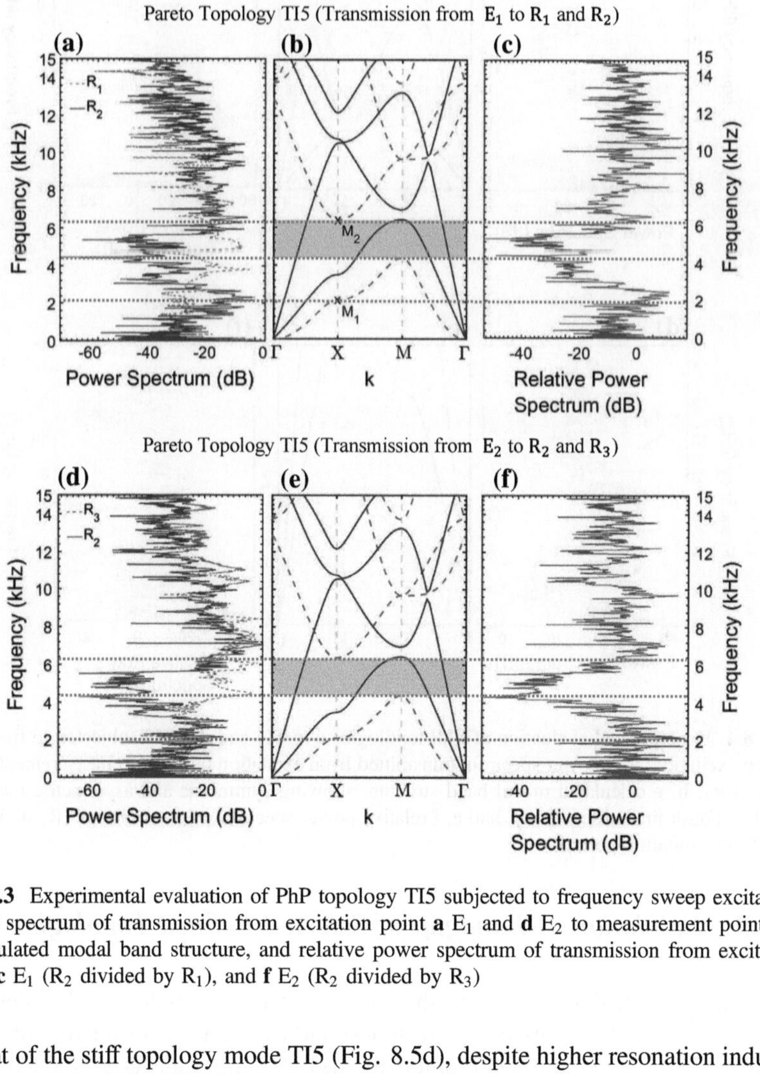

Fig. 8.3 Experimental evaluation of PhP topology TI5 subjected to frequency sweep excitation: power spectrum of transmission from excitation point **a** E_1 and **d** E_2 to measurement points; **b**, **e** calculated modal band structure, and relative power spectrum of transmission from excitation point **c** E_1 (R_2 divided by R_1), and **f** E_2 (R_2 divided by R_3)

to that of the stiff topology mode TI5 (Fig. 8.5d), despite higher resonation induced by the topology TI4 (Fig. 8.5a) which naturally intensifies the excitation. Based on the results given in Fig. 8.5 the stiff topology TI5 has slightly lower RBW but wider bandgap width with respect to the compliant topology TI4.

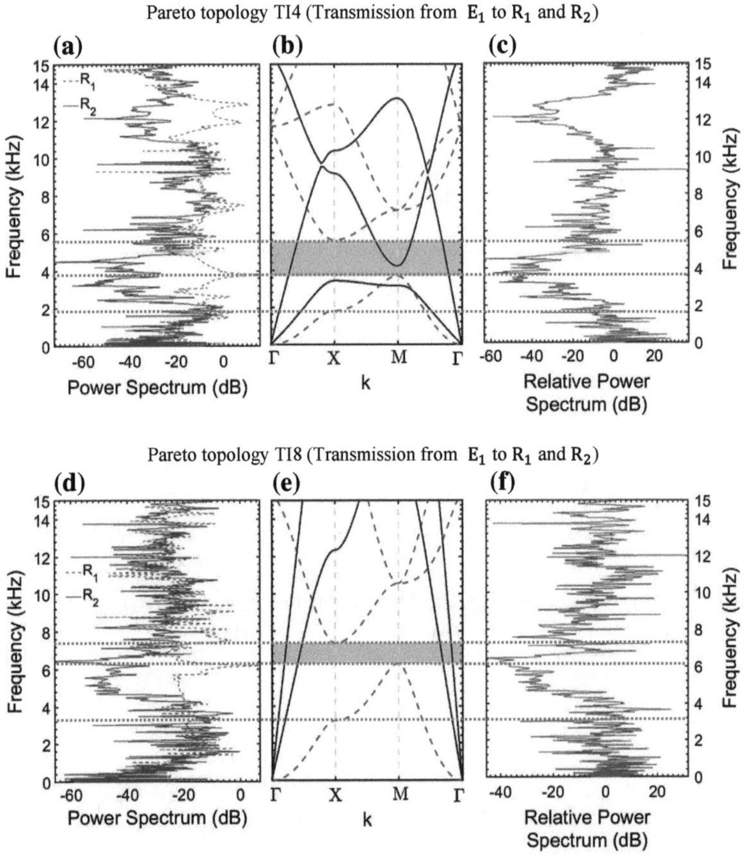

Fig. 8.4 Experimental evaluation of PhP topologies **a–c** TI4 and **d–f** TI8 subjected to frequency sweep excitation; **a**, **d** power spectrum transmitted from excitation point E_1 to measurement points R_1 and R_2; **b**, **e** calculated modal band structure showing symmetric and asymmetric modes by solid and dash lines respectively, and **c**, **f** relative power spectrum of transmission (R_2 divided by R_1) from excitation point E_1

8.2.2 Validation of Bandgap of Selected Refined Topologies

PhP plate of refined topologies, as shown in Fig. 8.6a for topology TJ3, is designed such that it can also be used for tensile test and experimental evaluation of elastic properties. An array of 7 × 7 unit-cells each 50 mm wide is perforated in an aluminium plate with thickness 5 mm and the transmission of Gaussian white noise from excitation point E_3 to measurement points R_4 and R_5 is determined. The contour of steady state transmission of vibrations excited at E_3 with bandgap frequency of 2.5 kHz and its attenuation throughout the designed PhP is shown in Fig. 8.6b for topology TJ3.

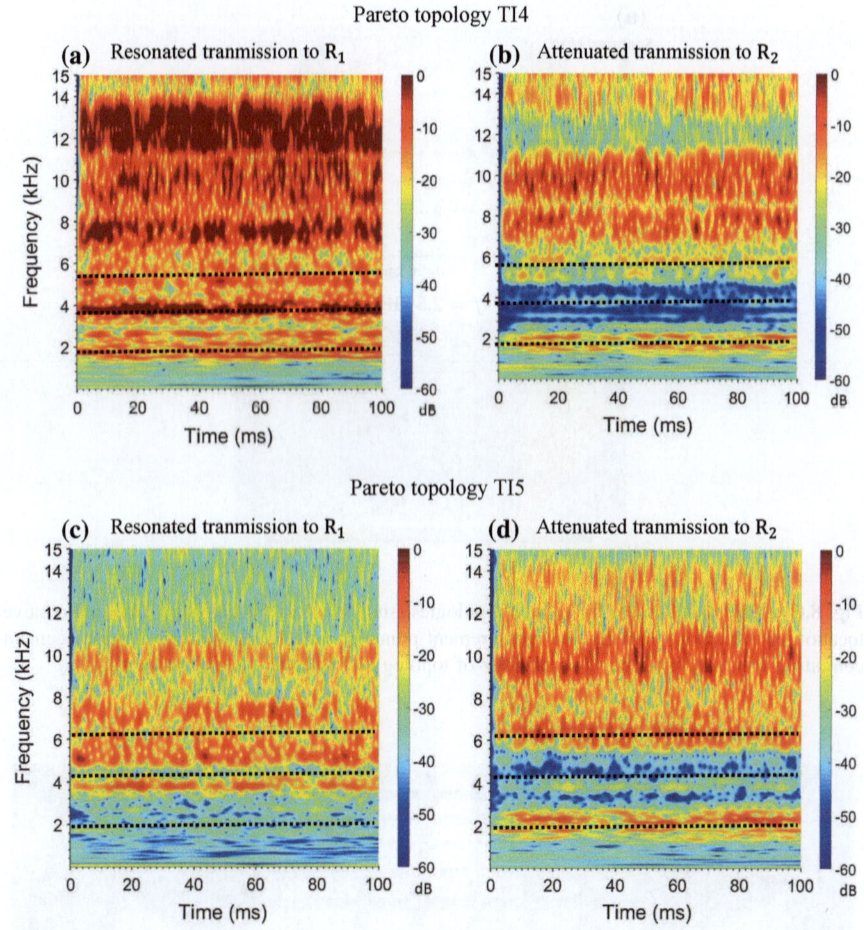

Fig. 8.5 Gaussian white noise excitation on E_1 and Gabor wavelet frequency-time spectrums of measured **a**, **c** resonated transmission to R_1 and **b**, **d** attenuated transmission to R_2 for Pareto topologies **a**, **b** TI4 and **c**, **d** TI5

The produced PhP of topology TJ6 is shown in Fig. 8.7a and close view of water-jetted topologies TJ3, TJ6 and TJ8 are shown in Fig. 8.7b–d respectively. PZTs of diameter 30 mm are attached to the top surface to excite and to measure the signal, and the same acquisition card (NI9234) is used. A Gaussian white noise is applied and its transmission spectrum is calculated up to 15 kHz through FFT over time duration of $t = 20$ ms as shown in Fig. 8.8. The spectrum of both measurement points which are located on the uniform sections is affected by the resonation of plate. So the applied white noise enables broad band excitation of frequencies and calculation of transmission spectrum over very short time to minimise the effect of resonation of boundaries. Spectra are normalised to the highest transmission amplitude measured at point R_5.

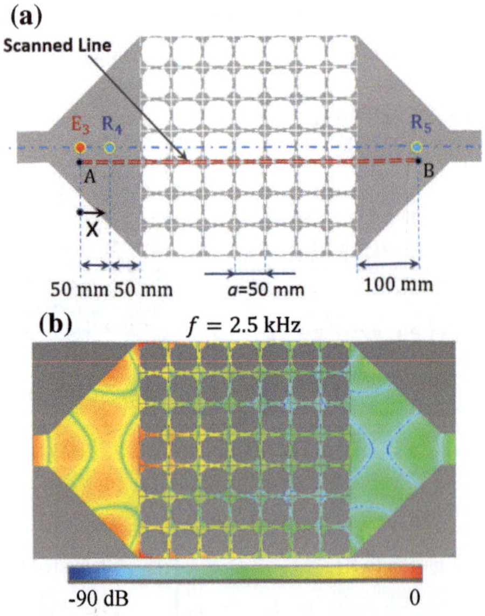

Fig. 8.6 **a** PhP plate design for refined topologies (shown herein for topology TJ3) and relative location of excitation point E_3 and measurement points R_3 and R_4, **b** steady state displacement contour at bandgap frequency $f = 2.5$ kHz of topology TJ3 applied on excitation point E_3

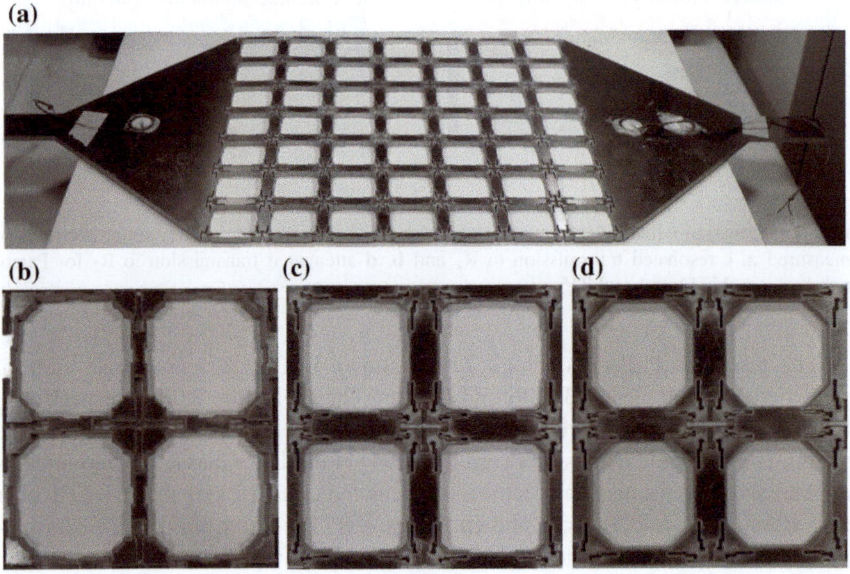

Fig. 8.7 **a** Produced PhP of refined topology TR6 according to Fig. 8.6a, and close view of produced topologies **b** TJ3, **c** TJ6 and **d** TJ8; (plate thickness $h = 5$ mm and unit-cell width $a = 50$ mm)

Measured transmission spectrums clearly confirm the calculated bandgaps of modal band structures as shown in Fig. 8.8 by a drop in transmission down to −40 dB. All spectrums start to drop at lower frequency level corresponding to opening partial bandgap of asymmetric wave modes along Brillouin zone border ΓX. Dominancy of this partial bandgap is expected for orthogonal transmission of wave through PhP from E_3 to R_5. The compliant topology TJ3 has an additional bandgap within measured frequency range at 11–14.5 kHz with significantly higher attenuation level down to −80 dB.

Comparing the modal band structure of topologies TJ6 and TJ8 (Fig. 8.8e, h) reveals that their partial bandgap along ΓX is almost the same while the complete bandgap of TJ8 is considerably narrowed by its lower modal branch.

Furthermore, a short 3-cycle sinusoidal tone-bust with central frequency tuned at around midgap frequency is applied at excitation point E_3 and its attenuation through successive PhP unit-cells is studied by scanning the line AB shown in Fig. 8.6a with a laser Doppler vibrometer (LDV). The normalised time-frequency spectrum of measured transmission to $x = 50$ mm (Fig. 8.6a) during measurement time $t = 5$ ms is calculated by Gabor wavelet transform as shown in Fig. 8.9a, c, e. The FFT of transmission to all points scanned over line AB is also calculated and the normalised contour of its variation with respect to x-axis (Fig. 8.6a) is shown in Fig. 8.9b, d, f; the white sections correspond to perforated parts of line AB. The dot lines highlight the frequency levels corresponding to complete and partial bandgaps, and the dash line shows the frequency level of applied excitation.

The Gabor wavelet spectra (Fig. 8.9a, c, e) demonstrate the actual spectrum of excitation and its variation versus time due to wave reflections from boundary and from porous PhP section. Initially, the spectrum has highest magnitude at around applied central excitation frequency 2.4, 3.0 and 3.5 kHz for topology TJ3, TJ6 and TJ8 (Fig. 8.9a, c, e) respectively. Then the spectrum is dominantly affected by the resonation of upper and lower modes confining the bandgap frequency. With regard to adequate excitation of complete bandgap frequency range, the displacement-frequency FFT contours on the right side of Fig. 8.9 confirm the attenuation of wave down to −60 dB within bandgap frequency after a few unit-cells.

The measurements of the topologies TJ3 and TJ6 are affected by an external low frequency noise source and a high pass filter is applied to the relevant FFT contours to remove the low level noise up to 300 Hz and clarify the measured excitation. The FFT contours show how the upper and lower modal branches of bandgap are resonated. The upper mode is strongly resonated and the lower mode is decayed after a few unit-cells as it is within the partial bandgap along Brillouin zone border ΓX.

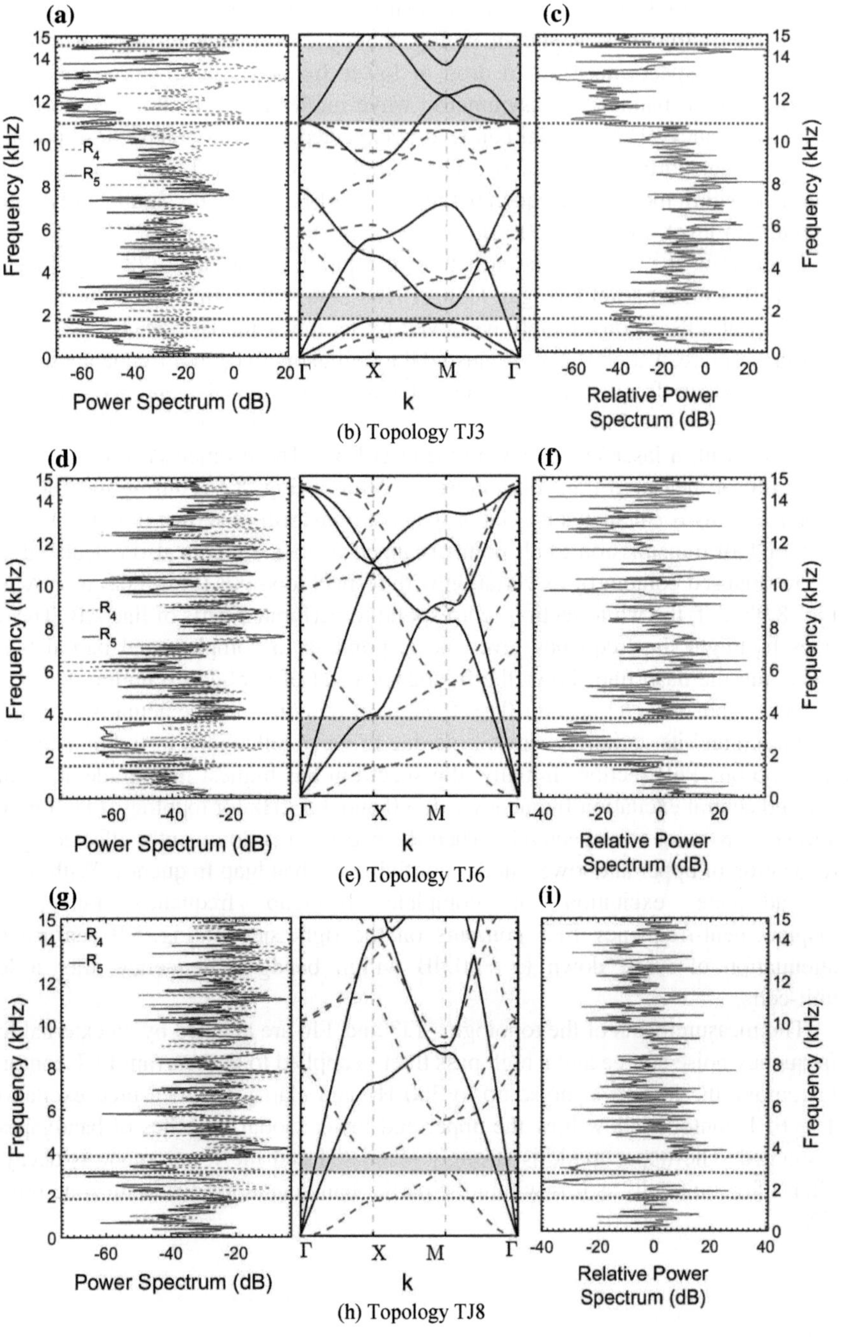

(b) Topology TJ3

(e) Topology TJ6

(h) Topology TJ8

◀**Fig. 8.8** Experimental evaluation of refined PhP topologies **a–c** TJ3, **d–f** TJ6 and **g–i** TJ8 subjected to Gaussian white noise excitation for duration of $t = 20$ ms at point E_3; **a**, **d**, **g** power spectrum of transmission to measurement points R_4 and R_5; **b**, **e**, **h** calculated modal band structure showing symmetric and asymmetric modes by solid and dash lines respectively, and **c**, **f**, **i** relative power spectrum of transmission (R_5 divided by R_4)

8.2.3 Validation of Stiffness of Selected Refined Topologies

The stiffness and effective elastic properties of optimised topologies, as the second objective of this multiobjective optimisation study, are also experimentally evaluated. Produced PhP specimens of refined topologies TJ3, TJ6 and TJ8, which are designed for this purpose, are stretched in the elastic regime by a tensile test machine as shown in Fig. 8.10a and the in-plane displacement of the central unit-cell (of 7 × 7 PhP array) is inspected.

The displacement of reference points M_{x1}, M_{x2} and M_{y1}, M_{y2} shown in Fig. 8.10b are measured during tensile loading to calculate the transversal and longitudinal strains of central unit-cell along x-axis and y-axis respectively. Quasi-static tensile tests (force controlled at a rate of 1 mm/min), are performed on an electromechanical 5800R tensile machine with 10 kN load cell. The maximum loading was taken sufficiently low to remain in the linear elastic region, and displacement over the central unit-cell was measured by (i) longitudinal and transverse extensometers and (ii) stereovision DIC.

In contrast to the use of extensometers, full field strain maps can be obtained by DIC (Sutton et al. 2009). This method is based on tracking the geometrical changes in the grey-scale distribution of a speckle pattern which is attached to the specimen surface. A random speckle pattern was applied on the specimens' surface by aerosol spray-paint and is monitored with 2 mega-pixel 8-bit CCD AVT Stingray F-201 B 1/1.8″ cameras throughout the loading event. During quasi-static loading, images with a 1624 × 1232 pixel resolution are acquired at a sampling rate of 2 Hz and are synchronised with the load–displacement signals from the tensile machine. A deformed image is taken at incremental loading steps and compared to the reference image taken prior to loading. This reference image is mapped by a square correlation subset window which is defined by the subset size and step size, being the pixel dimension of one single subset, and the centre distance between two adjacent subset windows, respectively. A correlation algorithm computes the displacement of each subset centre on the deformed image, estimating the displacement field across the region of interest (ROI). Determination of the specimen's surface displacement field was done using the VIC-3DTM software.

The contours of longitudinal displacement v and transversal displacement u in the central unit-cell of topologies TJ3, TJ6 and TJ8 measured by 3D-DIC are depicted in Fig. 8.11a–c respectively. The displacement fields correspond to a tensile load of 250.30, 756.47 and 753.92 N for TJ3, TJ6 and TJ8 respectively. Higher load level is required for the stiff topologies TJ6 and TJ8 for reliable

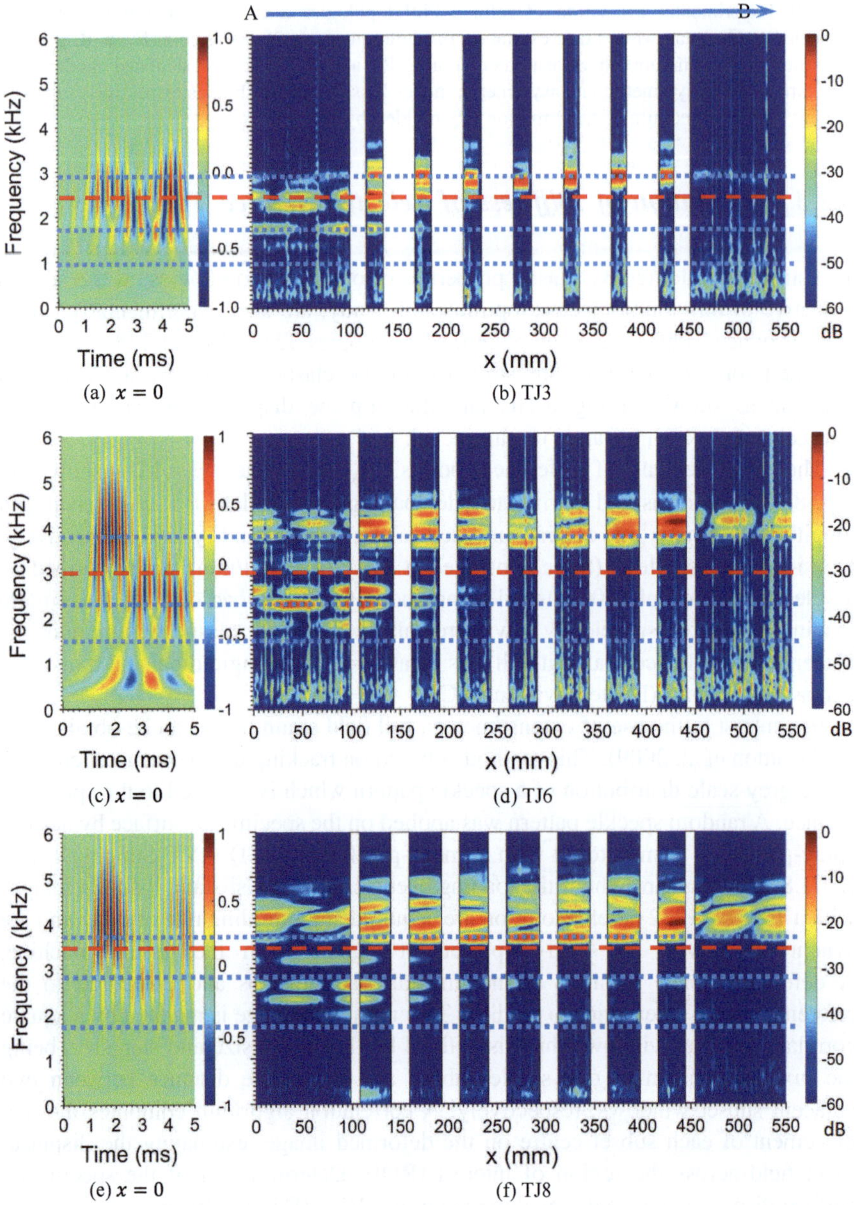

Fig. 8.9 Normalised transmission spectrums measured by LDV for 3-cycle sinusoid tone-burst applied at E_3, **a**, c, **e** wavelet spectrum of measured excitation at point A (Fig. 8.6a); **b**, **d**, **f** contour of spatial gradient of measured FFT over scanned line AB with length 550 mm as shown in Fig. 8.6a; **a**, **b** Topology TJ3 at tone-burst of 2.4 kHz, **c**, **d** Topology TJ6 at 3.0 kHz and **e**, **f** Topology TJ8 at 3.5 kHz

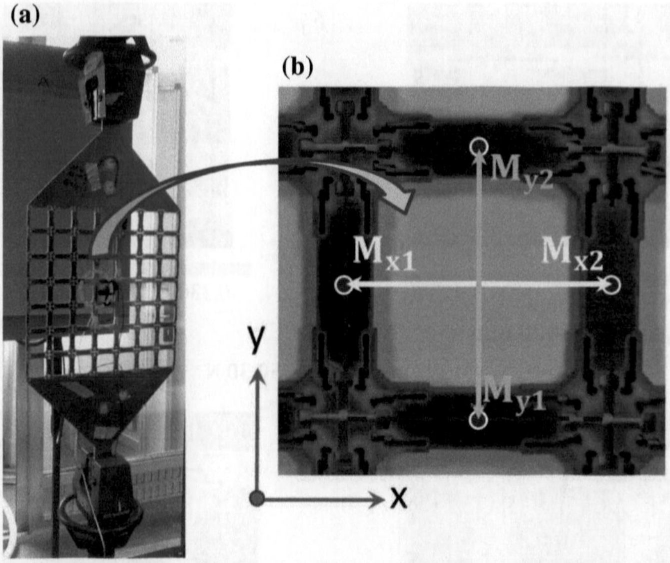

Fig. 8.10 **a** Tensile test and measuring the displacement field of central unit-cell, **b** measured reference points M_{y1}, M_{y2} and M_{x1}, M_{x2} to calculate the longitudinal and transversal strains of central unit-cell respectively (MMS Group, Ghent University)

measurement of induced displacements, and lower load level is necessary for the complaint topology TJ3 to ensure linear-elastic deformation of unit-cell.

From the measured displacement fields the expected relative compliance of topologies is evident. For a much lower applied load, topology TJ3 has much higher longitudinal deformation than TJ6 and TJ8. Due to symmetry of geometry and boundary conditions of PhP specimens, a uniform displacement field is ideally expected at the central unit-cell. However slight asymmetry is observable for topologies TJ6 and TJ8 (Fig. 8.11a–c) which might be due to misalignment of tensile load, rigid body motion of the specimen and also smaller deformations compared with TJ3 leading to increased measurement errors. The almost equal longitudinal extension and lateral contraction of topology TJ3 is evident in Fig. 8.11a.

The relations of normal stresses and strains for the plane-stress deformation of square symmetric PhP unit-cell based on its effective elastic properties are:

$$\varepsilon_x = \frac{\sigma_x - \nu_e \sigma_y}{E_e} \tag{8.2}$$

$$\varepsilon_y = \frac{\sigma_y - \nu_e \sigma_x}{E_e} \tag{8.3}$$

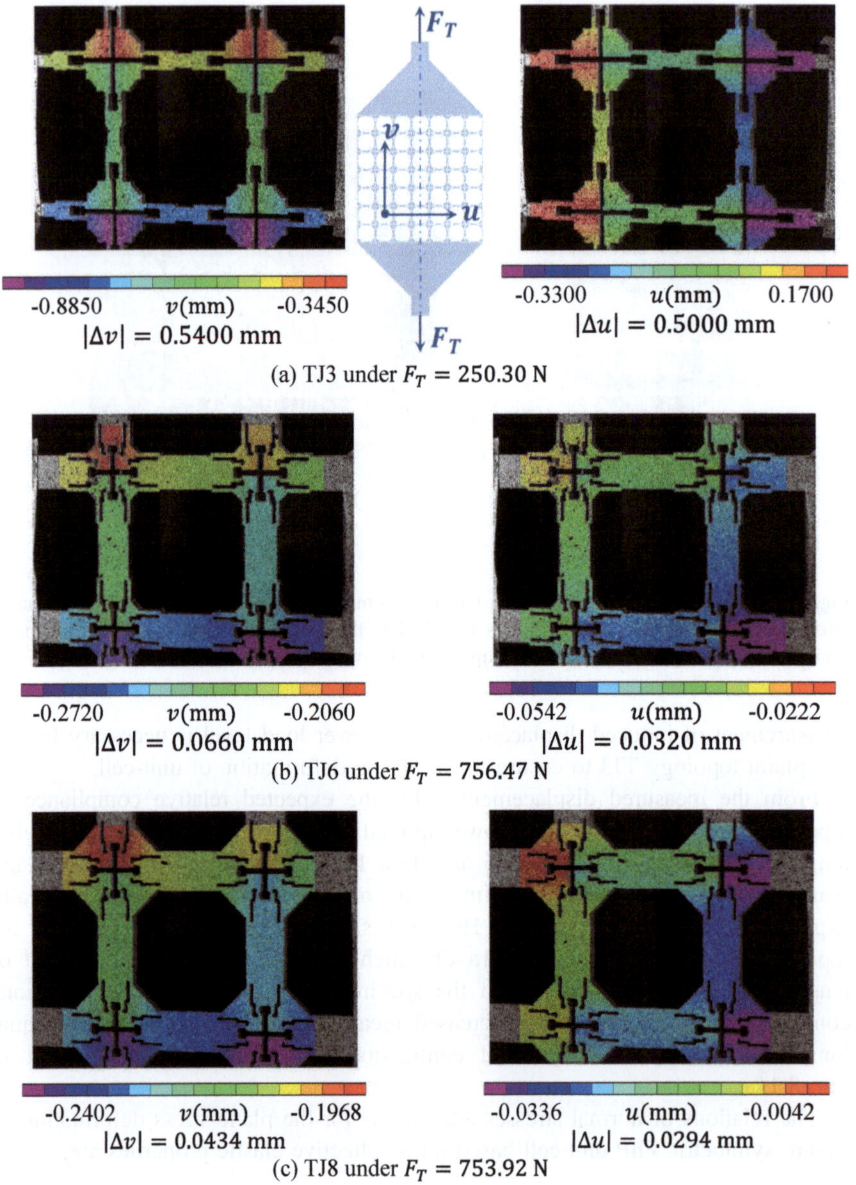

Fig. 8.11 Contour of displacement field (longitudinal v and transversal u) in the central PhP unit-cell measured by 3D-DIC, **a** TJ3 under tensile test load of $F_T = 250.30$ N, **b** TJ6 under $F_T = 756.47$ N and **c** TJ8 under $F_T = 753.92$ N

where the relevant small strains can be estimated based on the measured displacements:

$$\varepsilon_x = \frac{\partial u}{\partial x} = \frac{u_{Mx2} - u_{Mx1}}{a} \tag{8.4}$$

$$\varepsilon_y = \frac{\partial v}{\partial y} = \frac{v_{My2} - v_{My1}}{a} \tag{8.5}$$

where u_{Mx1}, u_{Mx2} and v_{My1}, v_{My2} are the transversal and longitudinal displacements of reference points (Fig. 8.10b) respectively. The effective stresses based on interfacial longitudinal load F_y and transversal load F_x of the unit-cell and for small deformation of unit-cell are:

$$\sigma_x = \frac{F_x}{ah} \tag{8.6}$$

$$\sigma_y = \frac{F_y}{ah} \tag{8.7}$$

Thus, the interfacial loads applied to the unit-cell are required for solving Eqs. (8.2) to (8.7) for E_e and v_e. In order to realise the deformation and loading conditions of central unit-cell, the tensile test is first simulated by ANSYS APDL (*ANSYS® Academic Research, Release 16*). An equal tensile load of $F_T = 70$ N is applied to PhP specimens of all topologies and resultant stress intensities and interfacial loads of central unit-cell are defined as shown in Fig. 8.12.

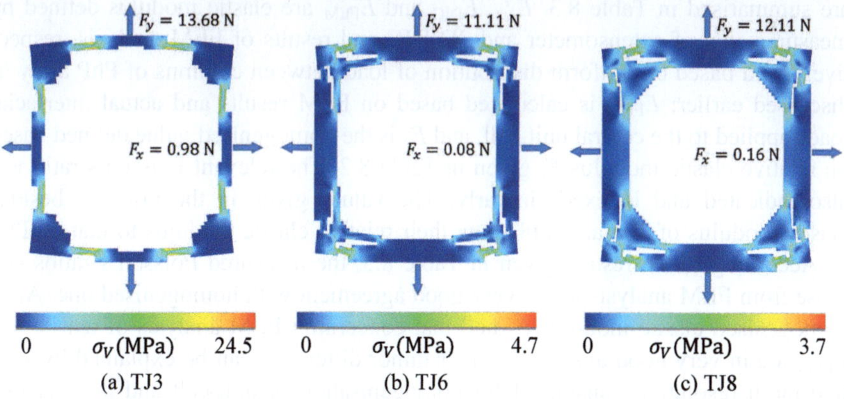

Fig. 8.12 FEM simulation of tensile test and resultant contour of von Mises stress σ_V in the central unit-cell and relevant interfacial loads, for tensile test load of $F_T = 70$ N (in average 10 N per column of unit-cells in the 7 × 7 PhP specimen as shown in Fig. 8.10a), for topologies **a** TJ3, **b** TJ6 and **c** TJ8, (Deformations are magnified for clarity)

The contour of von Mises stress σ_V shows how significantly the maximum stress in the unit-cell is reduced from 24.5 MPa for compliant topology TJ3 to 4.7 MP and 3.7 MPa for stiff topologies TJ6 and TJ8, respectively, under the same tensile load. For the applied tensile load $F_T = 70$ N, a 10 N load is expected per column of 7×7 PhP array by the assumption of uniform load distribution. However, as confirmed by the calculated longitudinal loads F_y depicted in Fig. 8.12, the actual longitudinal load applied to the central unit-cell is higher than uniform magnitude of 10 N. Moreover, a relatively small transversal load F_y is applied to the unit-cell (Fig. 8.12) slightly resisting its expected transversal contraction. Of course this transversal resistance is negligible for the stiff topologies TJ6 and TJ8. The non-uniform longitudinal load distribution and transversal loads are due to finite nature of specimen and the fact that transversal displacement of upper and lower boundaries of PhP section are constrained by the uniform plate sections having much higher stiffness.

In this research a uniform load distribution and no transversal load is assumed for estimation of elastic modulus and Poisson's ratio based on measured unit-cell displacements. However, the elastic properties are also calculated through FEM analysis of tensile test and both actual and uniform loading conditions are considered for comparison of relevant magnitudes with experimental measurements.

The tensile load is increased in the elastic regime and elastic properties are calculated based on relevant measurements. The results concerning measured elastic modulus and Poisson's ratio for compliant topology TJ3 and stiff topology TJ6 based on measurements of extensometer and 3D-DIC are depicted in Fig. 8.13. Regardless of the initial gradients due to inaccuracy of measuring very small displacements, linear response of the specimen under applied loading is obvious over which the elastic modulus and Poisson's ratio measured by extensometer and 3D-DIC converge and have good agreement.

All measured and analysed magnitudes of elastic modulus and Poisson's ratio are summarised in Table 8.3. $\bar{E}_{ext}$, $\bar{E}_{\mathrm{DIC}}$ and $\bar{E}_{\mathrm{FEM}}$ are elastic modulus defined by measurements of extensometer and 3D-DIC and results of FEM analysis, respectively, and based on uniform distribution of load between columns of PhP array as discussed earlier; E_{FEM} is calculated based on FEM results and actual interfacial loads applied to the central unit-cell, and E_{e} is the homogenised value defined based on relative elastic modulus E_{r} given in Table 8.2. The relevant Poisson's ratio v is also indicated and indexed similarly. The values given in the brackets besides elastic modulus of TJ6 and TJ8 show their relative elastic modulus to that of TJ3.

According to the results given in Table 8.3, the measured Poisson's ratios and those from FEM analysis are in very good agreement with homogenised one. Also, homogenised elastic modulus E_{e} and that concerning FEM analysis of tensile test E_{FEM} are in very good agreement; their minor difference can be explained by very fine mesh resolution considered for homogenisation of unit-cell and the fact that just two reference points are used in calculation of unit-cell strain for E_{FEM}. As expected, the $\bar{E}_{\mathrm{FEM}}$ defined based on uniform load distribution is underestimated because the averaged uniform load is less than actual load on the central unit-cell

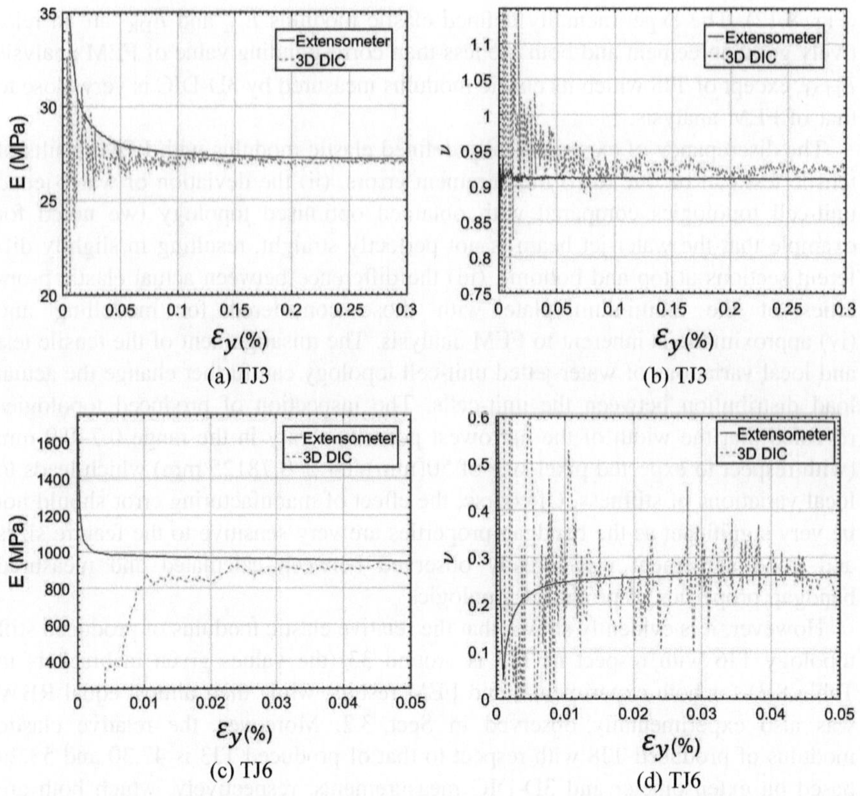

Fig. 8.13 Tensile test results measured by 3D-DIC and extensometer demonstrating **a**, **c** the measured elastic modulus and **b**, **d** Poisson's ratio versus longitudinal strain in the central PhP unit-cell, for topology **a**, **b** TJ3 and **c**, **d** TJ6

Table 8.3 Comparison of measured and analysed magnitudes of elastic modulus E and Poisson's ratio ν

Topology	TJ3	TJ6	TJ8
$\bar{E}_{ext}$ (MPa)	27.25 ± 0.03	905.0 ± 5.1 (33.21)	1288.8 ± 3.3 (47.30)
$\bar{E}_{\mathrm{DIC}}$ (MPa)	26.96 ± 0.16	909.1 ± 35.3 (33.72)	1436.7 ± 66.1 (53.29)
$\bar{E}_{\mathrm{FEM}}$ (MPa)	33.88	1129.17 (33.33)	1468.19 (43.34)
E_{FEM} (MPa)	43.29	1253.66 (28.96)	1763.59 (40.74)
E_{e} (MPa)	44.38	1266.16 (28.53)	1823.78 (41.09)
$\bar{\nu}_{ext}$	0.9109 ± 0.0003	0.2567 ± 0.0011	0.6361 ± 0.0012
$\bar{\nu}_{\mathrm{DIC}}$	0.9254 ± 0.0054	0.2693 ± 0.0498	0.6814 ± 0.0567
$\bar{\nu}_{\mathrm{FEM}}$	0.9101	0.2555	0.6175
ν_{FEM}	0.9217	0.2619	0.6256
ν_{e}	0.9198	0.2654	0.5897

The brackets include relative elastic modulus of topologies TJ6 or TJ8 with respect to TJ3

(Fig. 8.12). The experimentally defined elastic modulus $\bar{E}_{ext}$ and $\bar{E}_{\mathrm{DIC}}$ are in relatively good agreement and both are less than corresponding value of FEM analysis $\bar{E}_{\mathrm{FEM}}$, except of TJ8 which its elastic modulus measured by 3D-DIC is very close to that of FEM analysis.

The discrepancy of experimentally defined elastic modulus with FEM results of tensile test can be due to (i) measurement errors, (ii) the deviation of water-jetted unit-cell topologies compared with obtained optimised topology (we noted for example that the water-jet beam is not perfectly straight, resulting in slightly different sections at top and bottom), (iii) the difference between actual elastic properties of the aluminium plate with those considered for modelling and (iv) approximations inherent to FEM analysis. The misalignment of the tensile test and local variations of water-jetted unit-cell topology can further change the actual load distribution between the unit-cells. The inspection of produced topologies revealed that the width of the narrowest porosities vary in the range 0.7–0.9 mm (with respect to expected pixel size of $50(\mathrm{mm})/64 = 0.78125$ mm) which leads to local variations of stiffness. Of course, the effect of manufacturing error should not be very significant as the bandgap properties are very sensitive to the feature sizes and good agreement was already observed between calculated and measured bandgap properties of produced topologies.

However, it is evidently shown that the relative elastic modulus of produced stiff topology TJ6 with respect to TJ3 is around 33 (the values given in brackets in Table 8.3) for both experimental and FEM results, while their almost equal RBW was also experimentally observed in Sect. 3.2. Moreover, the relative elastic modulus of produced TJ8 with respect to that of produced TJ3 is 47.30 and 53.29 based on extensometer and 3D-DIC measurements, respectively, which both are higher than that of FEM calculations 43.34, while the produced topology TJ8 also has confirmed bandgap efficiency compared with its calculated RBW. Consequently, the relative effective stiffness and bandgap efficiency of topologies are experimentally validated as expected from the obtained Pareto front (Fig. 6.2c).

8.3 Optimised PhPs for Complete Bandgap of Mixed Guided Wave Modes

In this section experimental evaluation of produced PhP topologies with complete bandgap of mixed wave modes is presented and discussed. Intermediate Pareto topologies TM2 and TM5 (Fig. 6.7) with dominant RBW and effective stiffness respectively, are manufactured through water-jetting aluminium plates, and their calculated bandgap efficiency is experimentally validated. Moreover the topology TM6 (Fig. 6.7) is produced by laser cutting of a PMMA plate and its application for generating collimated wave at specified frequency above bandgap is investigated and demonstrated.

8.3.1 Water-Jetted Aluminium PhPs

Finite aluminium plates of thickness 2.5 mm and dimensions 500 × 400 mm are perforated by water-jetting equipment to make a central lattice array of 10 × 8 PhP cells with lattice periodicity of 25 mm as schematically shown in Fig. 8.14a for topology TM2. Orthogonal or diagonal dominant wave transmission from points E_4 and E_5 respectively, to the measurement point R_6 can then be measured. The transmission of signal from E_4 within the uniform plate and its resonation by PhP cells can be also measured at point R_6. Slots are cut on the border of designed PhP plate to isolate excitation segments and preferably prevent the excitation to reach the whole uniform border. The steady state transmission contour of an orthogonal excitation at E_4 at bandgap frequency 15 kHz for PhP of topology TM2 is presented in Fig. 8.14b in logarithmic scale dB. The contour clarifies exponential wave attenuation throughout the PhP lattice, wave resonation at excitation segment and also presents how the slots prevent transmission of wave to the whole border.

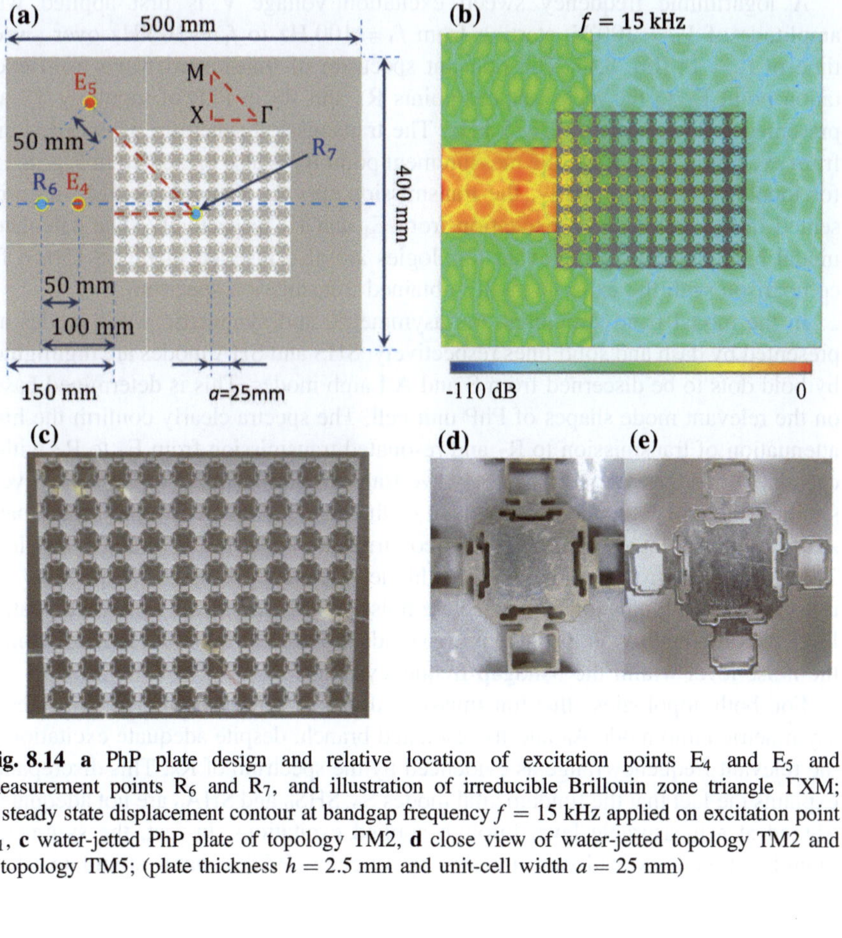

Fig. 8.14 **a** PhP plate design and relative location of excitation points E_4 and E_5 and measurement points R_6 and R_7, and illustration of irreducible Brillouin zone triangle ΓXM; **b** steady state displacement contour at bandgap frequency $f = 15$ kHz applied on excitation point E_1, **c** water-jetted PhP plate of topology TM2, **d** close view of water-jetted topology TM2 and **e** topology TM5; (plate thickness $h = 2.5$ mm and unit-cell width $a = 25$ mm)

Water-jetted PhP plate of topology TM2 and close view of water-jetted PhP cell of topology TM2 and TM5 are respectively shown in Fig. 8.14c, d, e.

Circular PZTs bonded to the top surface of plates with Salol are used to excite and measure guided waves. PZT with diameter of 20 mm is used for excitation and PZTs with diameter of 10 mm, which also fit on the central solid section of PhP unit-cell at R_7, are used for measuring transmitted signals. The four corners of the rectangular PhP specimens are suspended on small pieces of foam to prevent coupling of wave into the test table.

The measured signals are then recorded with a high resolution 24-bit acquisition card (NI9234) at a sampling frequency of 50 kHz in order to obtain an adequate evaluation of the wave amplitude decay for low frequency bandgap (up to 25 kHz). The wave signals are also recorded by an oscilloscope with a 16-bit acquisition card at a sampling frequency of 1 MHz. This allows to investigate transmission properties in a wider frequency range (up to 250 kHz). The frequency spectra are then calculated using fast Fourier transformation (FFT) and are normalised to the highest transmission amplitude measured at point R_2. The obtained spectra have been averaged over 20 measurements in order to improve the signal-to-noise ratio.

A logarithmic frequency sweep excitation voltage V is first applied with amplitude of $V_0 = 100$ V, starting from $f_0 = 100$ Hz to $f_1 = 30$ kHz over sweep time of $t_1 = 1$s (Eq. 8.1). The relevant spectrum of transmitted wave from excitation point E_1 to the measurement points R_1 and R_2 in PhP of topology T2 are presented in Fig. 8.15a up to 25 kHz. The transmission spectrum of exited signal from excitation point E_5 to the measurement point R_2 is also presented in Fig. 8.15c for topology TM2. Similarly, the transmission spectra of topology TM5 are presented in Fig. 8.15d, f for excitation from E_4 and E_5 respectively. The calculated modal band structure of relevant topologies are also inserted in Fig. 8.15b, d for comparison with the experimentally obtained transmission spectrum.

In the modal band structures, the asymmetric and symmetric wave modes are presented by dash and solid lines respectively. SHS and SHA modes are highlighted by bold dots to be discerned from S and A Lamb modes. This is determined based on the relevant mode shapes of PhP unit-cell. The spectra clearly confirm the high attenuation of transmission to R_7 and resonated transmission from E_4 to R_6 within calculated bandgap frequency. The wave transmission from E_5 to R_7 is also very similar to that of E_4 to R_6. Comparison of the bandgap efficiency and modal band structure of topologies TM2 and TM5 confirms the lower RBW of stiffer topology TM5 due to narrower bandwidth at higher frequency range. In all cases the transmission within bandgap reached the noise level of test setup. Better attenuation level of compliant topology TM2 is also evident due to its quite flat transmission at the noise level within the bandgap frequency range.

For both topologies, the transmission drops at lower frequency just above asymmetric lamb mode A_0 and its resonated branch, despite adequate excitation in the relevant frequency range as evidenced by the spectrum of R_6. This discrepancy explains the fact that the fundamental modes S_0, SHS_0 and SHA_0 are not adequately trigged at a measurable level through applied excitations. In fact, the symmetric omnidirectional excitations induced by circular PZT cannot stimulate shear

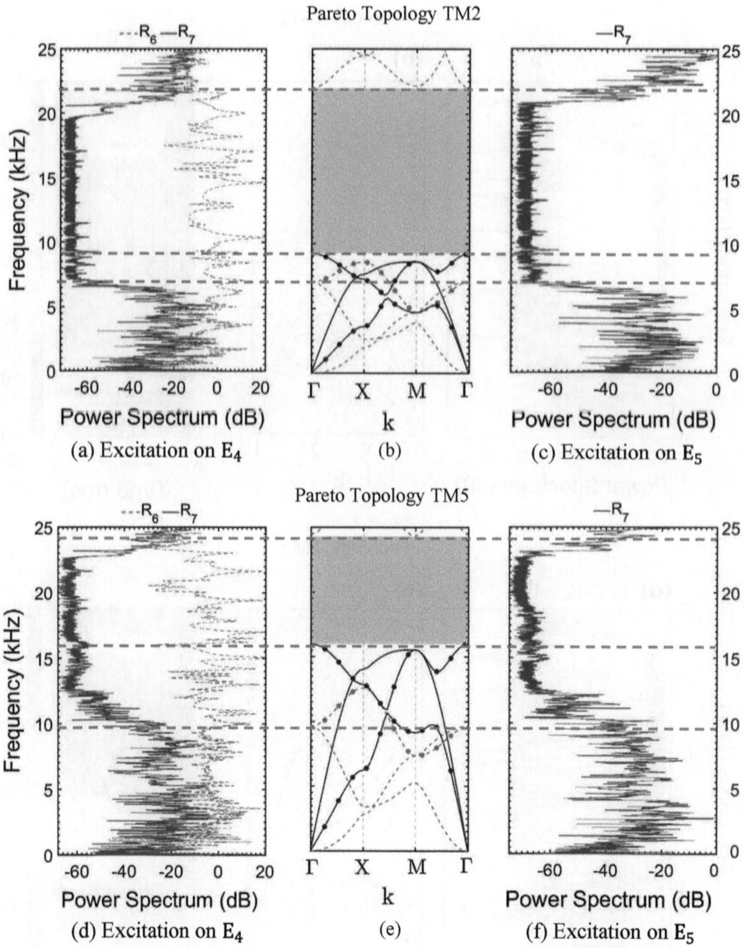

Fig. 8.15 Experimental evaluation of PhP topologies **a–c** TM2 and **d–f** TM5 subjected to frequency sweep excitation; **a**, **d** power spectrum transmitted from excitation point E_4 to measurement points R_6 and R_7; **b**, **e** calculated modal band structure, and **c**, **f** power spectrum transmitted from excitation point E_5 to measurement points R_7

horizontal modes and particular PZTs are needed for this purpose, e.g. square PZT design introduced by Zhou et al. (2015). As for the S_0 mode, the shortest wavelength corresponding to the frequency 25 kHz (with calculated phase velocity of 5413 m/s in the uniform aluminium plate of thickness 2.5 mm) is 216 mm which is much larger than the attached PZT dimeter of 20 mm. Thus the S_0 mode may not be excited at a measurable level in such a situation (Giurgiutiu 2003).

In order to investigate the competency of PhP specimens in the time domain, Gaussian white noise with amplitude of $V_0 = 100$ V is applied further as the excitation signal. Such random excitation enables simultaneous broad band

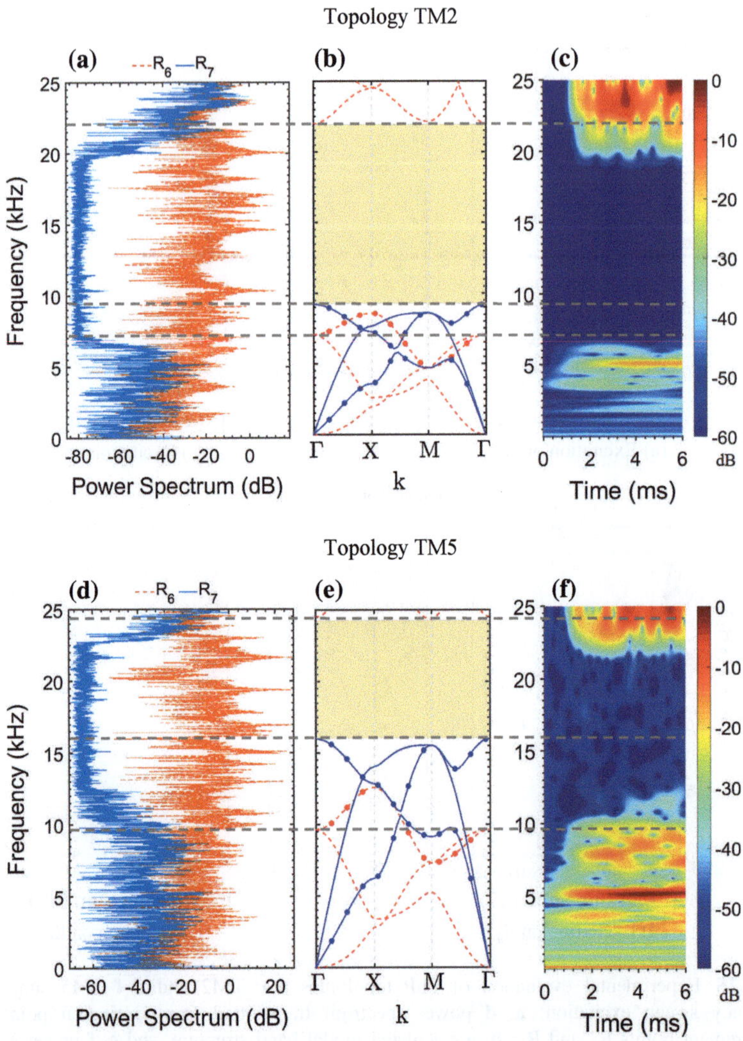

Fig. 8.16 Experimental evaluation of PhP topologies **a–c** TM2 and **d–f** TM5 subjected to Gaussian white noise excitation at excitation point E_4; **a**, **d** power spectrum of measured transmission to measurement points R_6 and R_7 up to 25 kHz, **b**, **e** calculated modal band structure, and **c**, **f** normalised Gabor wavelet spectrum of measured transmission to R_7 during $t = 6$ ms

excitation of guided waves with uniform energy distribution over the frequency domain. The overall power spectrum and also the Gabor wavelet time-frequency spectra of transmitted wave during $t = 100$ ms are then calculated, as shown in Fig. 8.16.

The Gabor wavelet transform reveals the time domain gradient of spectrum. All spectra are normalised by maximum transmission to the measurement point R_7, and wavelet spectrum contours are limited to −60 dB for more clarity. The same trend of transmission attenuation as obtained for frequency sweep excitation (Fig. 8.15) is observable for white noise excitation. The Gabor wavelet transmission spectra given in the Fig. 8.16c, f for topologies TM2 and TM5 respectively, specifically present the arrival of excited wave and its attenuation in measurement points R_7 during 6 ms interval. The relatively lower attenuation of transmission within bandgap of stiff topology TM2 (see Fig. 8.16f) is clear.

Long term wavelet spectra are further presented in Fig. 8.17 for both measurement points R_6 and R_7 during $t = 100$ ms. The resonated transmission of wave from PhP to the point R_6 and steady state bandgap efficiency of PhPs within

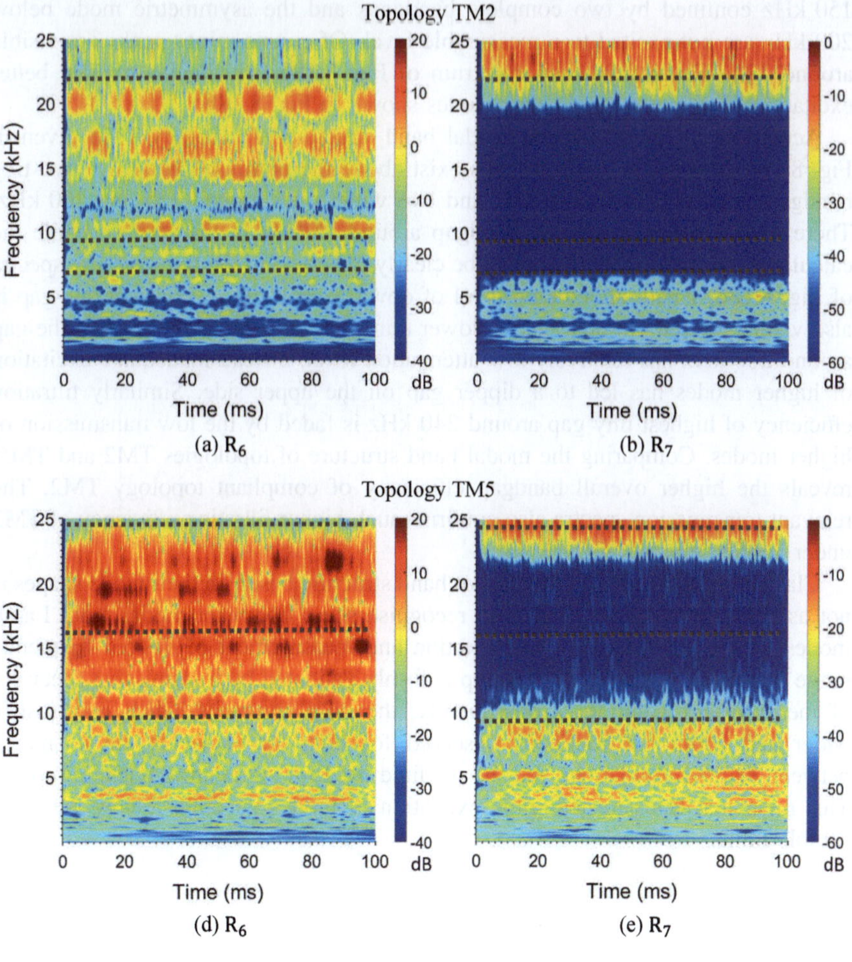

(a) R_6 (b) R_7 (d) R_6 (e) R_7

Fig. 8.17 Normalised Gabor wavelet spectra of transmitted Gaussian white noise from excitation point E_4 to the measurement points **a**, **d** R_6 and **b**, **e** R_7 in PhP plate of Pareto topology TM2 (top) and Pareto topology TM5 (bottom) over excitation time of $t = 100$ ms bandgap are fully confirmed. It is worth mentioning that wavelet spectrum contours of measurement points R_6 and R_7 have different limits as both have been normalised to the maximum transmission to R_7.

Broadband transmission spectra of excited white noise up to 250 kHz are also defined and evaluated with modal band structures of relevant topologies. As for the results concerning topology TM2 presented in Fig. 8.18a–c, the dispersion curves reveal four additional bandgaps above the optimised bandgap, at around 60, 100, 160 and 230 kHz. The obtained spectra (Fig. 8.18a, c) indicate a drop in transmission of down to −60 dB, confirming the presence of additional bandgaps. However, despite adequate broadband excitation observed by transmission spectrum of point R_7 (Fig. 8.18a), frequency ranges other than calculated complete bandgaps are also attenuated. Accordingly, the bunch of modes within 100–150 kHz confined by two complete bandgaps and the asymmetric mode below 200 kHz are not excited to a measurable level. Of course minor peaks are visible around 125 kHz in frequency spectrum of Fig. 8.18a, which may explain better excitation of relevant asymmetric modes shown in Fig. 8.18b.

According to the broadband modal band structure of topology TM5 given in Fig. 8.18e, four additional bandgaps exist above the optimised bandgap; two tiny bandgaps around 55 and 240 kHz and two wider ones around 135 and 200 kHz. There is a minor trace of tiny bandgap around 55 kHz in the spectra, while the calculated gap around 135 kHz can be clearly observed in the transmission spectra of Fig. 8.18d, f with attenuation level of down to −60 dB. A considerable gap is also visible around 165 kHz due to lower transmission of relevant modes. The gap around 200 kHz has relatively low attenuation level, instead inadequate excitation of higher modes has led to a dipper gap on the upper side. Similarly filtration efficiency of highest tiny gap around 240 kHz is faded by the low transmission of higher modes. Comparing the modal band structure of topologies TM2 and TM5, reveals the higher overall bandgap efficiency of compliant topology TM2. The relevant transmission spectra also confirm much higher filtration efficiency of TM2 under applied excitation.

Classifying the high order modes of band structure based on their mode shapes is not as easy as low order modes, to recognise shear dominant modes from Lamb modes. However, the adequate excitation and measurement of desired frequency range through implemented test setup is firmly validated by transmission spectrum of measurement point R_6. Consequently, the only argument which may justify wider and additional bandgaps observed for both topologies is the mismatch between excited oscillations and the required field for stimulating relevant modes. This mismatch leads to negligible excitation of modes generally below the measurable limits.

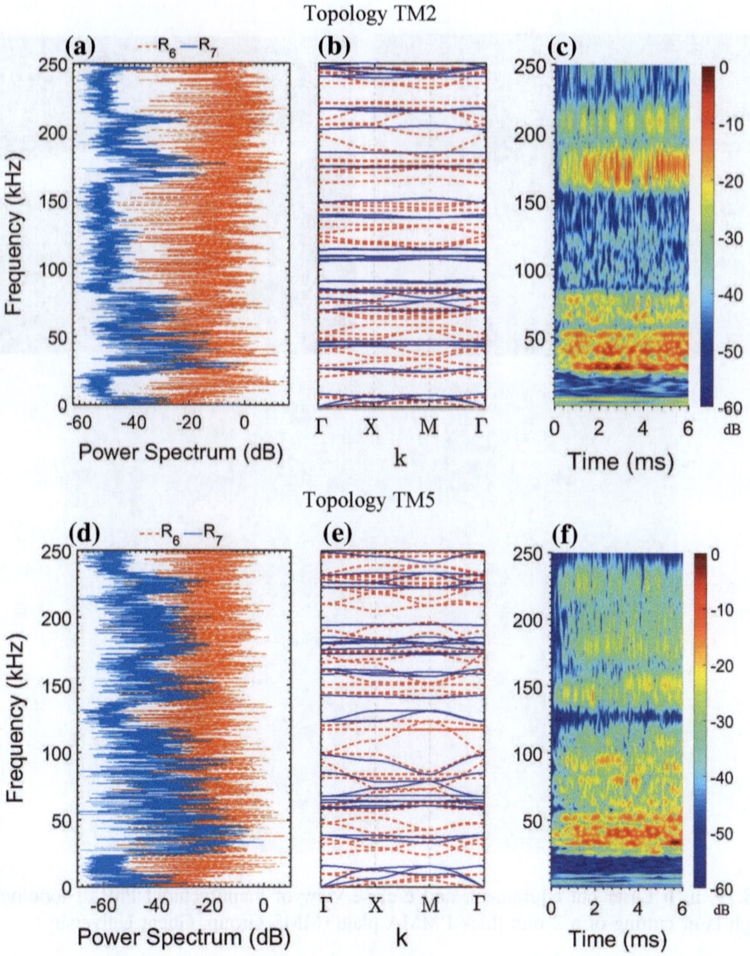

Fig. 8.18 Experimental evaluation of PhP topologies **a–c** TM2 and **d–f** TM5 subjected to Gaussian white noise excitation at excitation point E_4; **a, d** power spectrum of measured transmission to measurement points R_6 and R_7 up to 250 kHz, **b, e** calculated modal band structure, and **c, f** normalised Gabor wavelet spectrum of measured transmission to R_2 during $t = 6$ ms

8.3.2 Laser Cut PMMA PhP

In order to investigate the wave steering functionality of achieved porous PhPs, Pareto topology TM6 (Fig. 6.7) is alternatively selected and manufactured through laser cutting of a PMMA plate as shown in Fig. 8.19. PMMA plate with thickness 2 mm, mass density $\rho_s = 1180\ \mathrm{kg/m^3}$, Poisson's ratio $\nu_s = 0.34$ and dynamic elastic modulus $E_s = 6.5$ GPa is used.

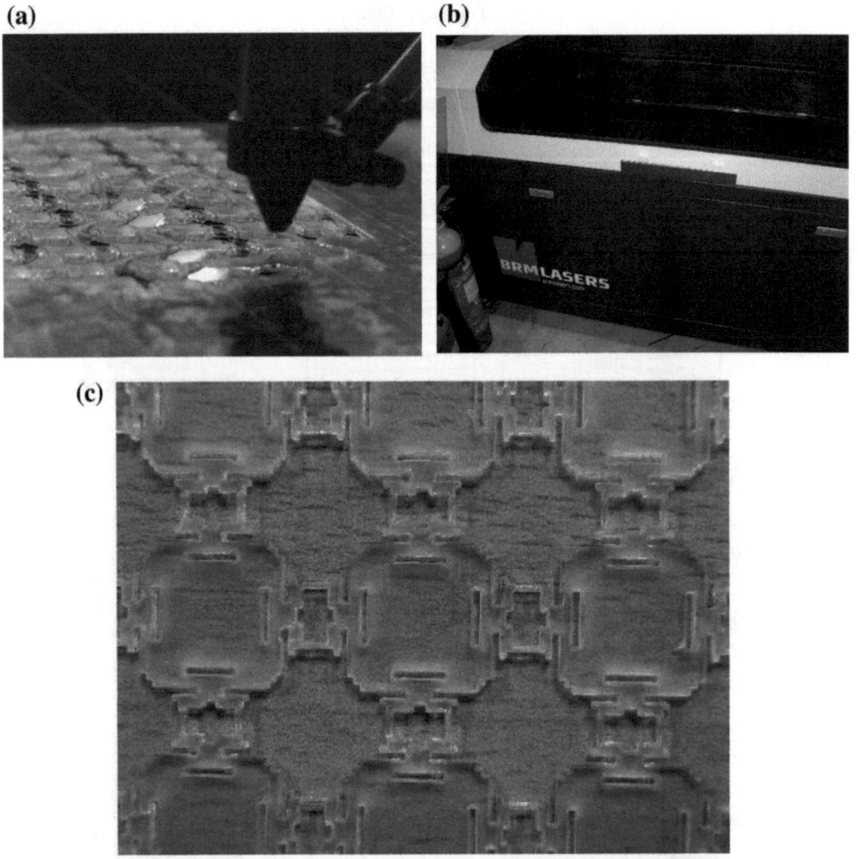

Fig. 8.19 **a**, **b** Laser cut equipment, and **c** close view of manufactured PhP of topology TM6 through laser cutting of a 2 mm thick PMMA plate (MMS Group, Ghent University)

The calculated modal band structure of T6 with PMMA base material is shown in Fig. 8.20a including a complete bandgap in the frequency range 9.2–12.8 kHz. The focus of this portion of research is to study the possibility of using selected PhP topology for self-collimation at frequencies above bandgap through asymmetric wave modes A and B marked in Fig. 8.20a. For this purpose the nature of these modal branches is first analysed by extracting a few mode shapes corresponding to e.g. ΓX border of Brillouin zone. Mode shapes of points M_1 and M_3 at Γ with wave vector $\mathbf{k} = \{0 \quad 0\}$, and points M_2 and M_4 at X with wave vector $\mathbf{k} = \{\pi/a \quad 0\}$ are calculated. The side view of relevant mode shapes along x-axis and in zx-plane are shown in Fig. 8.20b presenting the normalised contour of transversal displacement w in z-axis. The mode shapes of both modes at Γ (i.e. $\mathbf{k} = \{0 \quad 0\}$) reveal a complete wavelength ζ along the unit-cell width a in x-axis.

Consequently, modes A and B are resonated modes folded back into the first Brillion zone under prescribed periodic boundary condition, and the mode shapes of

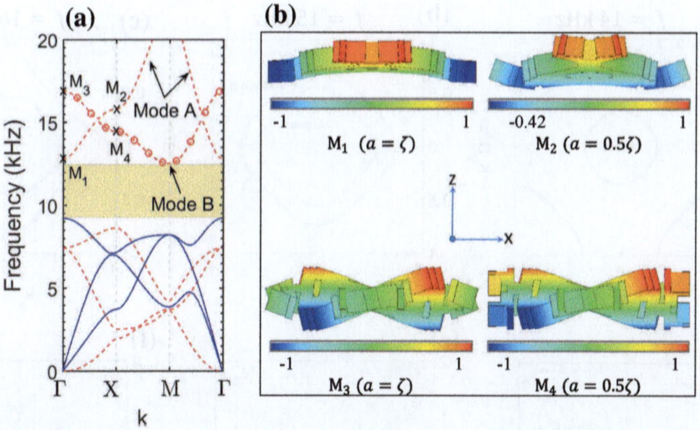

Fig. 8.20 **a** Modal band structure of Pareto topology TM6 for unit-cell width $a = 20$ mm and PMMA base material, **b** side view of mode shapes corresponding to the points M_1, M_2, M_3 and M_4 marked on the modal band structure concerning Brillouin zone border ΓX (Fig. 3.4a) and relevant wavelength ζ compared to the unit-cell width a; contours show normalised transversal displacement in z-axis

M_1 and M_3 correspond to the actual wave vector $\mathbf{k} = \{2\pi/a \quad 2\pi/a\}$ outside first Brillouin zone. The mode shapes also explain the dominant bending and torsional deformation nature of asymmetric modes A and B respectively. So it is expected that symmetric omnidirectional excitation induced by a circular PZT, dominantly excite bending mode A.

The EFCs of the modes A and B at frequencies 14, 15 and 16 kHz by considering Bloch-Floquet periodic boundary condition are first calculated as shown in Fig. 8.21a–c inside Brillouin zone. As already shown in Fig. 8.20b, these modes are folded into the first Brillouin zone. To clarify, the actual unfolded EFCs are also presented in Fig. 8.21d–f with thick solid lines, and their periodicity in the PhP lattice is highlighted through thin dash lines. Basically the shape of EFC profile is the same as it is linearly mapped from first Brillouin zone space $\mathbf{k} \in [0, \pi/a]^2$ into a higher space $\mathbf{k} \in [\pi/a, 2\pi/a]^2$ taking into account the fact that shifting from 0 to π/a corresponds to $2\pi/a$ to π/a. However, the difference is in the interpretation of EFCs. For example, the EFCs in the actual wave vector space given in Fig. 8.21e show almost concave wave front of both modes at 15 kHz while the Fig. 8.21b is opposite. Of course the central flat section remains the same at both presentations.

It is well known that the vector of group velocity in lattice xy-plane at specified frequency ω can be defined as $\mathbf{C_g} = \{\partial\omega/\partial k_x \quad \partial\omega/\partial k_y\}$ which is normal to the EFC. According to Fig. 8.21 the wave propagation characteristic of topology TM6 is highly sensitive to the frequency around 15 kHz. At 14 kHz mode A with concave EFC is converging and mode B with convex EFC is diverging (Fig. 8.21d), and vice versa at 16 kHz (Fig. 8.21f). But, at 15 kHz both modes are converging with almost concave EFCs and a flat section in the middle (Fig. 8.21e).

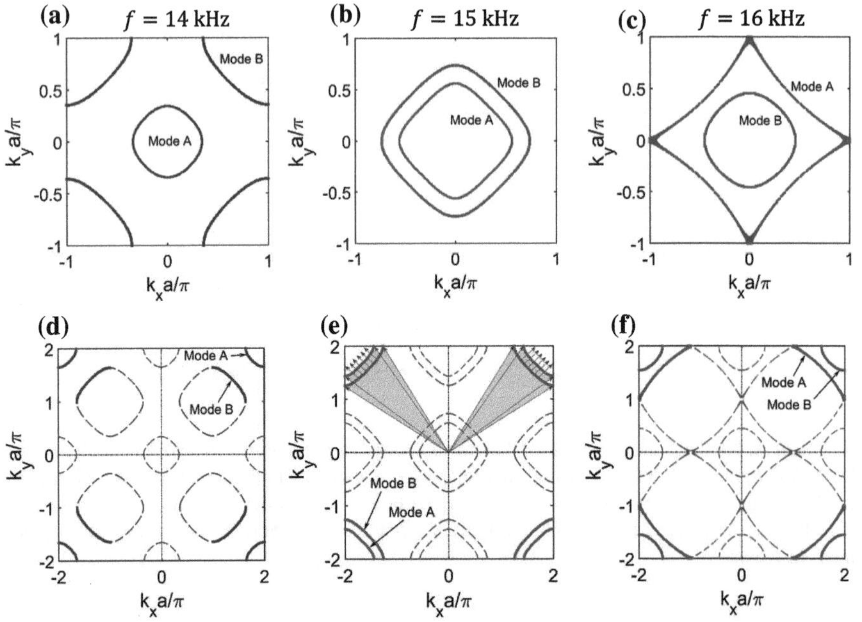

Fig. 8.21 Calculated EFCs of topology TM6 with PMMA base material at frequencies **a**, **d** 14 kHz, **b**, **e** 15 kHz and **c**, **f** 16 kHz on the upper side of bandgap; **a–c** folded periodic EFC in Brillouin zone, **d–f** actual EFC outside Brillouin zone (solid lines) and its periodicity in PhP lattice (dash lines)

These flat sections of the EFCs stand for flat wave fronts which confine the energy flow along the normal vectors as shown in Fig. 8.21e for the upper half space.

In order to investigate the functionality of calculated EFCs of topology TM6, a 15×6 array of this topology, each of $a = 20$ mm wide, is perforated in a rectangular plate of dimensions 600×420 mm and thickness 2 mm, as shown in Fig. 8.22a. The produced PhP plate is presented in Fig. 8.22b. The middle-bottom cell of the PhP array is left unperforated to attach a PZT of 20 mm diameter for excitation (Fig. 8.22a). A 3-cycle sinusoid windowed with Hanning function is then applied at central frequency of 15 kHz for dominant excitation of this frequency. Then the rectangular area above the PhP section shown in Fig. 8.22b is scanned by LDV. The transmitted signal is recorded in a regular xy-grid by the scanning table shown in Fig. 8.23, and the snapshots of recorded amplitude are defined over the whole scanned area to visualise relevant wave front. An acoustic foam panel is used to insulate the specimen from the scanning table (Fig. 8.23). The snap shots of wave transmission throughout scanned area after 600, 700 and 750 μs from excitation time are presented in Fig. 8.24.

The snapshots clearly show arrival and propagation of two rays of wave with concave wave front into the uniform plate area corresponding to the EFC calculated at frequency 15 kHz shown in Fig. 8.24e. The two asymmetric modes A and B

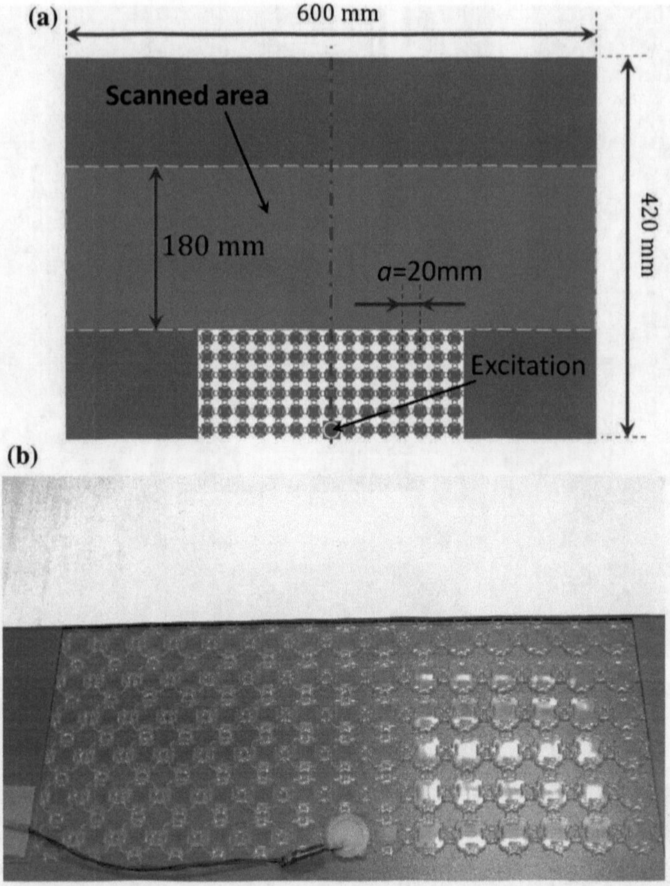

Fig. 8.22 **a** Designed PhP plate of Pareto topology TM6 and dimensions; and **b** produced PhP plate through laser cutting of a PMMA plate of thickness 2 mm with reflective tape attached to the area to be scanned by LDV

shown in the EFC of 15 kHz (Fig. 8.24e) are parallel, and have slightly different wave vectors, therefore propagation speeds. Thus the distinction of modes A and B from the propagated wave front throughout scanned area is practically difficult. However, as mentioned earlier, dominant excitation of Mode A with bending deformation (Fig. 8.22b) is expected through implemented PZTs. Although the applied tone-burst includes dominant excitation of waves with central frequency of 15 kHz, other frequencies around 15 kHz with significantly different EFCs (Fig. 8.21) are also excited and contribute to the observed wave front. Nonetheless, the snapshots of the measured wave front are in good agreement with the calculated EFC at 15 kHz.

◂**Fig. 8.23** Test set up and scanning table for measurement of the wave propagation throughout the laser cut PMMA specimen by LDV; the acoustic foam panel is used to prevent coupling of excited waves with the scanning table (MMS Group, Ghent University)

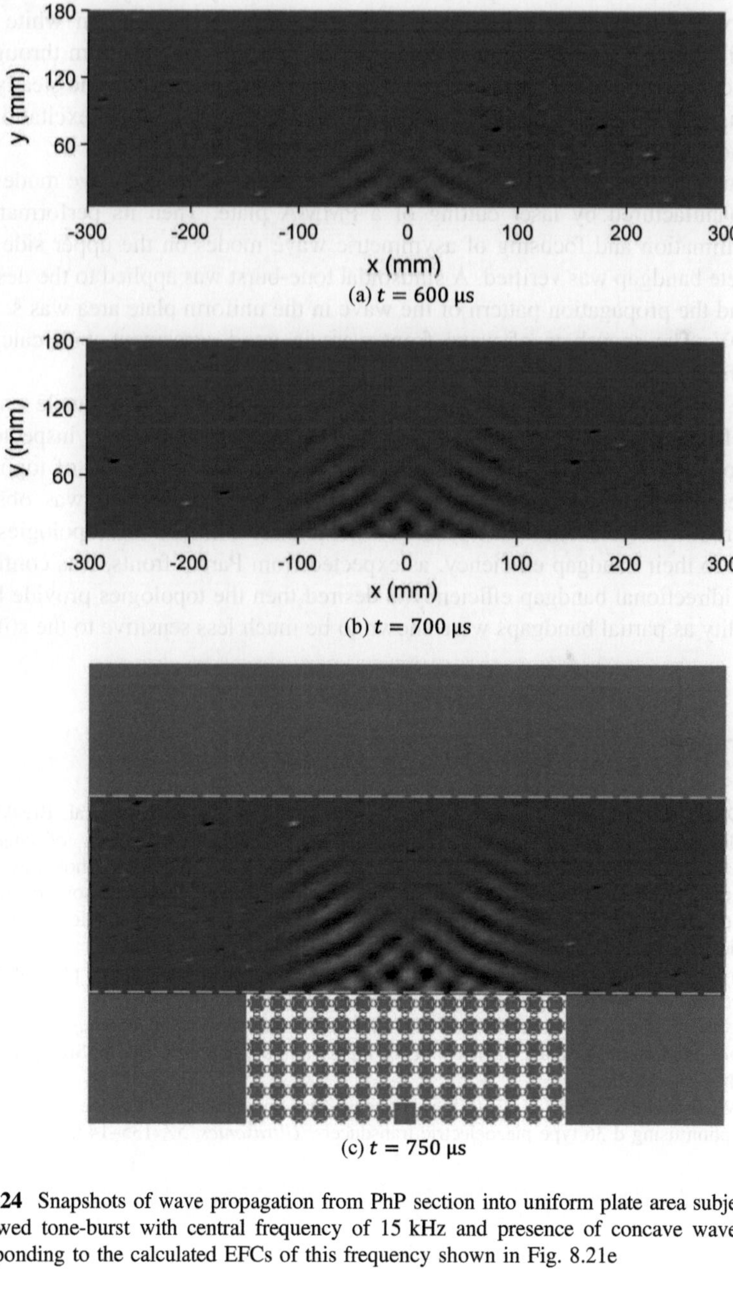

Fig. 8.24 Snapshots of wave propagation from PhP section into uniform plate area subjected to windowed tone-burst with central frequency of 15 kHz and presence of concave wave fronts corresponding to the calculated EFCs of this frequency shown in Fig. 8.21e

8.4 Concluding Remarks

In this chapter the experimental validation of selected optimised porous PhPs, using the results presented in Chap. 6, and manufactured by water-jetting aluminium plates was presented and discussed. Frequency sweep and random white noise excitations were applied and the attenuation of transmission spectrum through the PhP section within calculated bandgap frequencies was observed. The steady state bandgap efficiency of finite PhPs subjected to random white noise excitation was also shown through Gabor wavelet spectrums of transmitted signal.

Also an optimised topology, with complete bandgap of mixed wave modes, was also manufactured by laser cutting of a PMMA plate. Then its performance in self-collimation and focusing of asymmetric wave modes on the upper side of its complete bandgap was verified. A sinusoidal tone-burst was applied to the designed PhP and the propagation pattern of the wave in the uniform plate area was scanned by LDV. The snapshots of wave front were in good agreement with calculated equal frequency contours of the relevant topology.

The stiffness of selected specimens were also evaluated through tensile tests and the deformation of the central unit-cell of phononic lattice was inspected by extensometer and 3D-DIC. The elastic modulus and Poisson's ratio of topologies were estimated based on the measurements and good agreement was observed compared with relevant FEM results. The relative stiffness of topologies with respect to their bandgap efficiency, as expected from Pareto fronts, was confirmed. If a unidirectional bandgap efficiency is desired then the topologies provide higher flexibility as partial bandgaps were shown to be much less sensitive to the stiffness.

References

Bilal, O. R., & Hussein, M. I. (2012). Topologically evolved phononic material: Breaking the world record in band gap size. In *Photonic and phononic properties of engineered nanostructures* (pp. 826911–826917). International Society for Optics and Photonics.

Giurgiutiu, V. (2003). Lamb wave generation with piezoelectric wafer active sensors for structural health monitoring. In *Smart structures and materials* (pp. 111–122). International Society for Optics and Photonics.

Halkjær, S., Sigmund, O., & Jensen, J. S. (2006). Maximizing band gaps in plate structures. *Structural and Multidisciplinary Optimization, 32*(4), 263–275.

Sutton, M. A., Orteu, J. J., & Schreier, H. (2009). *Image correlation for shape, motion and deformation measurements: Basic concepts, theory and applications*. Berlin: Springer Science & Business Media.

Zhou, W., Li, H., & Yuan, F.-G. (2015). Fundamental understanding of wave generation and reception using d 36 type piezoelectric transducers. *Ultrasonics, 57,* 135–143.

Chapter 9
Conclusions and Recommendations for Future Work

9.1 Introduction

Extraordinary features of heterogeneous lattice structures in controlling acoustic and elastodynamic waves have attracted a great deal of research to develop their applications, necessitating their design optimisation. The destructive interaction of waves with lattice structures when the wavelength is comparable to the lattice periodicity, enables self-collimation, negative refraction and even total reflection of particular frequencies. Successive in-phase (Bragg) reflection of waves at the interface of periodic heterogeneities causes exponential decay of the wave's amplitude within phononic bandgap frequency.

Guided waves produced through thin-walled plate like structures travel long distances with low loss. Hence phononic crystal plates (PhPs) can be designed for production of low loss acoustic devices e.g. resonators, wave guides and filters. Moreover, PhPs can be designed to perform as a built-in acoustic device and manipulate guided waves for health monitoring of large plate like structures. Symmetric and asymmetric Lamb modes, and symmetric and asymmetric shear horizontal wave modes are the well-known guided wave modes. These modes have different dispersion properties and also interact differently with various defect types. So it is desirable to manipulate wave modes through exclusive bandgaps of symmetric or asymmetric wave modes. A complete bandgap of mixed wave modes is also desirable for e.g. designing plate structure with broadband vibration isolation or total reflection and resonation of wave energy.

The present research was particularly dedicated to topology optimisation of 1D and 2D PhPs for manipulation of guided waves and in this context novel research contributions were made in relation to designing and optimising phononic crystals in general. Porous 2D PhPs were extensively studied and selected optimised topologies were also manufactured and experimentally validated. The main contributions and conclusions of this research are summarised below.

S. Hedayatrasa, *Design Optimisation and Validation of Phononic Crystal Plates for Manipulation of Elastodynamic Guided Waves*, Springer Theses,
https://doi.org/10.1007/978-3-319-72959-6_9

9.2 Conclusions

9.2.1 1D Bi-Material Layered PhPs

Regarding 1D PhPs bi-material layered design was assumed. Maximised RBW between lowest possible couple of Lamb wave modal branches, and also maximised total RBW of first 11 Lamb modal branches were studied. The gradient of optimised topology with respect to filling fraction of constituents was studied and compared, for symmetric and asymmetric unit-cell designs, and following results were achieved:

- Lowest complete bandgap of Lamb waves in 1D bi-material PhPs was opened between 3rd and 4th modal branches. The unit-cell topology for maximised RBW of this bandgap converged to the regular topology with centric scattering inclusion for a wide range of filling fractions. However, for the lower range of filling fractions (of scattering inclusion) the RBW was significantly increased compared to the regular centric topology. This filling fraction range with increased RBW was slightly widened as the aspect ratio (width to thickness) of unit-cell was increased from 2 to 4. Relaxing the symmetry of topology didn't offer topologies with noticeably better bandgap efficiency. It was shown that maximising RBW of Lamb waves degrades the complete bandgap of plate waves due to presence of a shear horizontal interrupting mode.
- Total RBW of Lamb waves between first 11 modal branches in 1D bi-material PhPs of aspect ratio 2 was significantly increased for a large portion of filling fractions, compared to regular centric topology. Relaxing the symmetry of unit-cell led to significantly better topologies with increased total bandgap efficiency.
- A definite symmetric topology including two stiff scattering counter parts was introduced for various filling fractions and its lower bandgap and also total bandgap efficiency had excellent agreement with optimised symmetric topologies.
- It was shown that the bandgap efficiency of different gradient PhPs produced through consecutive optimised unit-cells of different filling fractions is almost the same as bandgap efficiency of a uniform PhP comprised of identical optimised unit-cells of the same average filling fraction. The difference of gradient and uniform PhPs in low frequency response due to having different fundamental natural frequencies was shown, highlighting the multiscale functionality of gradient PhPs.

9.2.2 *2D Porous PhPs and In-Plane Stiffness*

As for 2D PhPs, single material porous design with square symmetry was particularly studied. PhPs with porous heterogeneities, manufactured through perforation of a uniform plate, have perfect wave reflecting interfaces and are free from interfacial imperfections. However, optimisation of such porous PhPs for maximised bandgap efficiency naturally leads to a topology with maximally isolated solid domains i.e. weakest connections. Hence it is essential to obtain optimised PhPs with adequate stiffness. For this purpose, a multi-objective topology optimisation was carried out for maximised RBW and maximised in-plane stiffness, and the following key results were achieved:

- The vitality of taking into account the effective stiffness of topology when optimising PhPs for maximised RBW was confirmed. The substantial dependency of the bandgap efficiency of porous PhP (and generally porous Phononic Crystals) to the in-plane stiffness was presented. It was concluded that increasing the stiffness generally degrades the bandgap efficiency and vice versa. Topologies with almost identical bandgap efficiency but different stiffness modes (shear and/or normal) were achieved.
- The sensitivity of modal band structure of PhP unit-cell to compliant material properties commonly considered as a substitute of void elements was studied. It was shown that reducing the solid material properties, i.e. density and elastic modulus, by the order of magnitude of more than 5, used as compliant material, provides accurate results for modal band analysis of relevant porous unit-cell.
- Complete bandgaps of mixed wave modes and exclusive bandgaps of symmetric and asymmetric guided wave mode were maximised. There is no record in the literature in relation to optimised complete bandgap and exclusive bandgap of symmetric modes. However, optimised topologies for bandgaps of asymmetric wave modes indicated superior bandgap efficiency compared to preceding works available in the literature. Thick PhP with aspect ratio 2 generally provided higher RBW compared to thin PhP with aspect ratio 10.
- The topology refinement significantly shifted the Pareto front of bandgaps of asymmetric modes towards higher bandgap-stiffness efficiency. However, the Pareto front of complete bandgaps was shifted marginally. This is mainly due to having stricter bandgap objective in the case of complete bandgaps (to open bandgap between all mode types) which restricts the solutions space.
- In all optimisation cases, a sudden change in the structure of topology was observed, leading to a considerable gap i.e. discontinuity in the intermediate section of Pareto front. The topology mode introduced on the stiff side of discontinuous section of Pareto, provides almost the same bandgap properties of the other topology on the compliant side, but through considerably higher stiffness. For the case of bandgaps of asymmetric modes, the discontinuous section of Pareto front was widened through topology refinement, leading to increased stiffness contrast of relevant stiff and compliant topology modes.

- The variety of topologies obtained between stiff and compliant extremes of achieved Pareto fronts offer great choices for designing gradient phononic crystal plates in which spatial variation of stiffness is enabled while maintaining an almost uniform bandgap efficiency. Hierarchical arrangement of selected topologies also offer creation of very wide bandgaps while maintaining the overall stiffness at the maximised level.

9.2.3 2D Porous PhPs and Deformation-Induced Tunability

The optimisation of 2D porous PhPs was also extended to design of tunable/ switchable bandgaps. A multiobjective study was performed for (i) maximised total bandgap of mixed wave modes within first 12 modal branches and (ii) maximised or minimised gradient of total bandgap through large nonlinear deformation. Elastomeric hyperelastic solid material and unit-cell aspect ratio of 2 were assumed to be stretched up to 10% under equibiaxial tensile load. Novel topologies were obtained and following key conclusions were made:

- Optimised tunable porous PhPs for maximised deformation-induced RBW gradient introduced novel designs with degraded bandgap through deformation. All optimised topologies have maximised sensitivity to the prescribed deformation. An extreme topology has highest bandgap efficiency but not totally degraded through deformation. Another extreme topology has lower bandgap efficiency which completely closes through deformation.
- Optimised tunable porous PhPs for minimised deformation-induced RBW gradient also introduced novel designs with other promising tunability characteristics: (i) widest bandgaps with minimum sensitivity to external load, (ii) deformation-induced bandgaps which do not exist in initial state, and (iii) intermediate topologies with initial bandgap efficiency widened via external load.

9.2.4 Experimental Validation of Optimised Porous 2D PhPs

Critical topologies obtained for thin porous 2D PhPs with aspect ratio 10 were selected and experimentally validated. Various topologies were selected across Pareto fronts so that relatively compliant, stiff and very stiff topologies were included. Bandgap efficiency and effective stiffness of topologies manufactured by water-jetting aluminium plate and laser cutting PMMA plate were experimentally evaluated. The followings are the major conclusions drawn:

- (i) Frequency sweep, (ii) random Gaussian white noise and (iii) sinusoidal tone-burst excitations were applied and their transmission spectrums through produced PhPs were measured by laser Doppler vibrometer (LDV) and Piezoelectric transducers (PZTs). The measured transmission spectrums showed excellent bandgap efficiency of tested finite PhP specimens and full agreement was observed with calculated modal band structures of relevant PhP unit-cells under perfect periodic boundary conditions.
- The effective stiffness of PhPs of a few topologies were evaluated by a tensile test machine. The deformation of the central unit-cell of designed PhPs was measured by (i) mechanical extensometers and (ii) 3D stereovision. Effective elastic modulus and Poisson's ratio were calculated and compared with homogenised values and also results from FEM simulation of tensile test. With regards to the manufacturing imperfections and measurement errors, a good agreement was observed.
- As expected, the stiffer topology exhibited narrower bandgap at slightly higher frequency range with relatively lower attenuation level. Of course, for topologies with exclusive bandgap of asymmetric wave modes the unidirectional bandgap efficiencies were shown to be much less sensitive to the stiffness.
- Since the lower side of optimised complete bandgaps of mixed wave modes is confined by fundamental shear horizontal mode and symmetric Lamb wave modes, the implemented PZTs were not able to adequately excite them to a measurable level. So the measured bandgaps were wider, in the lower side, and expanded down to the frequency level corresponding to fundamental asymmetric Lamb wave mode, as confirmed by the modal band structure of relevant unit-cells calculated through FEM.
- The transmission spectrum of topologies with bandgaps of asymmetric wave modes indicated a considerable drop within calculated bandgap frequency which was extended down to the frequency level concerning relevant partial bandgap. Scanning the wave transmission throughout the specimens by LDV clearly demonstrated the attenuation of excited wave within bandgap frequency and resonation of bandgap side bands.
- The bandgaps of asymmetric wave modes were at a frequency level considerably lower than complete bandgap of mixed wave modes which enable manipulation of larger wavelengths through the same unit-cell size. However, the complete bandgaps of mixed wave modes were much wider and showed much higher bandgap attenuation efficiency.
- The equal frequency contours of modal branches on the upper side of bandgap were calculated for an arbitrary optimised topology. Flat and concave wave fronts were realised at particular frequency and the efficiency of produced PhP in collimation and focusing a tone-burst of that frequency were experimentally confirmed. The snapshots of wave front were in good agreement with calculated equal frequency contours of the relevant topology.

9.3 Recommendations for Future Work

The research presented in this thesis was concerned with topology optimisation of PhCrs in unit-cell scale. Layered bi-material 1D PhPs and square symmetric porous 2D PhPs with uniform through thickness perforation profile were studied. Stochastic optimisation GA was essentially employed to handle the disjoint objective function of total bandgap within multiple modal branches, and to handle the discontinuity of design domain and associated complications when optimising porous bandgaps.

Followings are potential research topics recommended for furthering this research:

- Optimisation of 1D PhPs was limited to bandgaps of Lamb waves, and optimisation of tunable 2D PhPs was limited to complete bandgap of mixed guided wave modes. It is therefore worthy to explore exclusive bandgaps of symmetric and asymmetric guided waves for 1D PhP and tunable 2D PhP designs.
- Other lattice shapes, e.g. hexagonal or rectangular, with or without prescribed symmetry may also be studied to investigate their impact on bandgap, stiffness and tunability efficiency and to introduce anisotropic performance where applicable.
- Based on the realised sensitivity of PhPs to the filling fraction and effective stiffness, it is worthwhile to develop a multiscale optimisation algorithm for designing gradient PhP structures with multiscale functionality.
- Using GA for stochastic optimisation of bandgap structures with a large number of design variables is impractical and computationally extremely intensive. However, preceding gradient based optimisation studies have been unsuccessful is terms of maximising multiple bandgaps and also in finding feasible porous bandgap topologies without pseudo intermediate domains. So it is worthwhile to develop a robust and computationally efficient optimisation algorithm for this purpose.

Curriculum Vitae

Saeid Hedayatrasa
Postdoctoral Fellow
Mechanics of Materials and Structures
Department of Materials, Textiles and Chemical Engineering
Ghent University
9052 Zwijnaarde, Belgium
e-mail: saeid.hedayatrasa@ugent.be
Scopus Author ID: 55845267000

Education	
2012–2016	PhD in Mechanical and Manufacturing Engineering University of South Australia, Adelaide, Australia
2000–2003	Master's Degree in Mechanical Engineering Tarbiat Modares University, Tehran, Iran
1996–2000	Bachelor's Degree in Mechanical Engineering University of Mazandaran, Mazandaran, Iran
Honors and Awards	
2017	Postdoctoral Research Grant by Ghent University
2017	Nomination for South Australian PhD Research Excellence Award
2016	Nomination for University of South Australia Ian Davey Thesis Prize
2015	International Mobility Travel Grant by University of South Australia
2015	Early Career Fellowship Award by 11th World Congress of Structural and Multidisciplinary Optimisation (WSMO-11), University of Sydney
2012	University President's PhD Scholarship by University of South Australia
Work Experience	
2013–2017	Tutor and Student Assessor, University of South Australia
2012	Mechanical Designer, Mobarakeh Steel Engineering Company (MSE)
2006–2012	Mechanical Designer, BIDEC Company, IRITEC Group

S. Hedayatrasa, *Design Optimisation and Validation of Phononic Crystal Plates for Manipulation of Elastodynamic Guided Waves*, Springer Theses,
https://doi.org/10.1007/978-3-319-72959-6

MIX
Papier aus verantwortungsvollen Quellen
Paper from responsible sources
FSC® C105338

If you have any concerns about our products,
you can contact us on
ProductSafety@springernature.com

In case Publisher is established outside the EU,
the EU authorized representative is:
Springer Nature Customer Service Center GmbH
Europaplatz 3, 69115 Heidelberg, Germany

Printed by Libri Plureos GmbH
in Hamburg, Germany